Hormones as Tokens of Selection

Qualitative Dynamics of Homeostasis and Regulation in Organismal Biology

Hugo van den Berg

CRC Press is an imprint of the
Taylor & Francis Group, an **informa** business

CRC Press
Taylor & Francis Group
6000 Broken Sound Parkway NW, Suite 300
Boca Raton, FL 33487-2742

Printed on acid-free paper

International Standard Book Number-13: 978-0-367-13441-9 (Hardback)

Library of Congress Cataloging-in-Publication Data

Names: Berg, Hugo van den, 1968- author.
Title: Hormones as tokens of selection : qualitative dynamics of homeostasis and regulation in organismal biology / Hugo Antonius van den Berg.
Description: Boca Raton, Florida : CRC Press, [2019] | Includes bibliographical references and index.
Identifiers: LCCN 2018050903| ISBN 9780367134419 (hardback : alk. paper) | ISBN 9780429026508 (ebook)
Subjects: | MESH: Hormones--genetics | Homeostasis--genetics | Biological Evolution | Models, Biological | Systems Biology
Classification: LCC QP571 | NLM WK 102 | DDC 612.4/05--dc23
LC record available at https://lccn.loc.gov/2018050903

Visit the Taylor & Francis Web site at
http://www.taylorandfrancis.com

and the CRC Press Web site at
http://www.crcpress.com

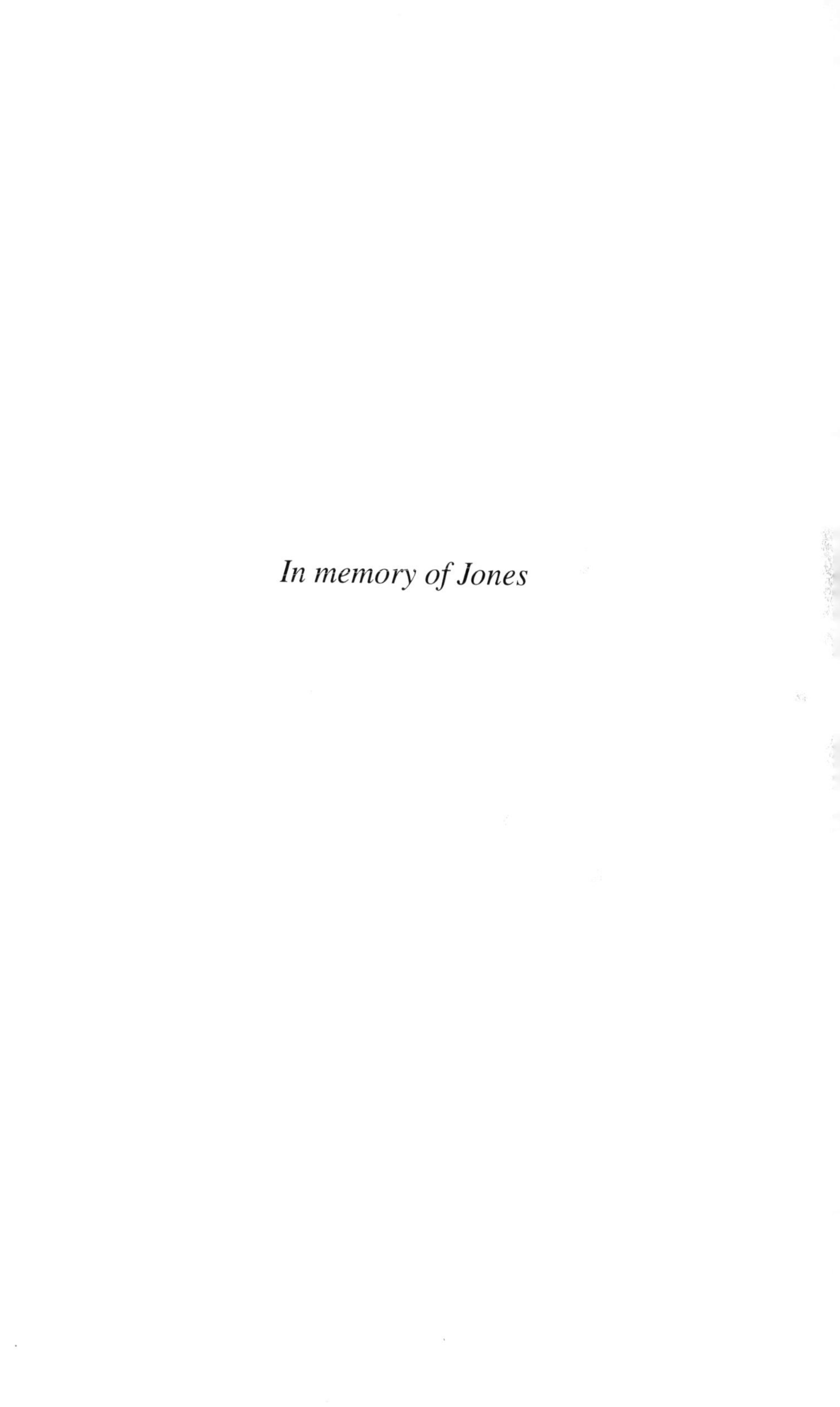

In memory of Jones

Contents

Author

Hugo van den Berg teaches mathematical biology at the University of Warwick. He has written several textbooks on evolutionary dynamics and on the general theory and practice of mathematical modelling, as well as research papers on energetics and homeostasis, T-cell immunity, evolutionary biology, bioinformatics and the physiology of uterine contractility.

CHAPTER 1

Introduction

Hormonal pleiotropy describes the intriguing phenomenon where each hormone affects a number of target tissues and processes, and in turn each such target is affected by multiple hormones [16]. The list of a given hormone's effect is known as the *suite* mediated by the hormone, or also the *suite of traits* affected by the hormone [43]. The task of learning these suites by heart is the bane of the student, who wonders why there could not simply be a single, dedicated messenger for each target process. Even when on occasion the hormone's name suggests just that, things are seldom quite what they seem; e.g., growth hormone (GH) is *not* 'the' growth hormone, a name that belongs more properly to Insulin-like Growth Factor-1 (IGF-1) [16].

One possible reply to the student's lament is sheer economy: with hundreds or even thousands of distinct targets, pleiotropy could drastically reduce the number of messenger species required. Examples of target processes include local rates of biosynthesis or degradation, product secretion, as well as differentiation and proliferation of specific cell types [16]. However, genomic real estate is unlikely to constitute a limiting factor in most metazoans, and therefore this argument fails to rule out convincingly the possibility of one-to-one matching of each messenger to a single dedicated target.

Alternatively, we might console and excite the student with the notion that networks with a high level of connectivity tend to be robust and adaptable. That is, should there occur, in evolutionary time, a shift in the demands that are made on regulation, then those regulatory systems that happen to be equipped with some built-in redundancy may be capable of being nudged in an evolutionarily favourable direction by fewer (or smaller) mutations. If such intrinsic pliability plays a major role, the systems we encounter in the field have not merely been selected because of the adaptive configuration of their internal control systems, but they are also present because they have effectively been meta-selected for 'evolvability.'

A third response hinges on the concept of *organismal harmony*, the attainment of integrated regulation at the level of the whole organism. This argument presents pleiotropy as the natural consequence of the unity of purpose that arises at the level of the whole organism through cooperation of its cells and tissues. On this view, a hormone's effect on various targets aligns the behaviour of these targets at the level of some common purpose that can be identified at the organismal level. At the same time, any given target's susceptibility to several hormones means that these purposes become entangled, giving rise to a qualitative dynamics of interlocking control loops that gives regulation in biological systems its special flavour. The aim of the present monograph is to develop these ideas.

1.1 HORMONAL PLEIOTROPY AND ORGANISMAL HARMONY

The argument from organismal harmony explains pleiotropy as a concomitant of unity of purpose at the level of the whole organism, and of hormones acting for these purposes. This idea is by no means novel—indeed, it frequently crops up in informal conversations among (neuro)endocrinologists and in seminar rooms, where loose talk of 'purpose' is less likely to face censure. If we are to give this idea its rightful place in biological theory, we should be able to endow 'unity' and 'purpose' with respectable meanings.

The unitary nature of the organism expresses itself precisely in the fact that certain choices are made at this level; for instance, fight versus flight (short time scales), hunt/forage versus rest/digest (intermediate time scales), and somatic growth versus reproduction (longer time scales). When we use terms such as 'choice' and 'decision' we simply mean that a path is picked from among several alternatives, without necessarily meaning to imply any grounding in conscious deliberation. All that is required is some sort of centralised process that integrates the available information or stimuli that bear on these decisions. Organismal harmony with respect to these decisions refers to the concerted activity across different organs and tissues, that is, local adjustments in accordance with the global decision.

Whenever a choice is made, there are repercussions for the organism's fitness. In general, the alternatives each come with both costs and benefits to fitness. This is often understood as bargaining the fitness that alternative A might have contributed against the fitness associated with alternative B [43, 101]. Such a choice can be regarded as a *trade-off*. It seems *prima facie* reasonable to suppose that the trade-offs achieved are balancing points at which fitness is maximised, although the application of optimisation *per se* to evolutionary scenarios is not without its perils [12]. For instance, developing and maintaining the information-processing apparatus carries its own fitness penalty: regulation has to earn its keep and this sets boundaries on how sophisticated the decision-making machinery can be.

The term *life history trade-off* is used primarily when discussing the intermediate-to-lifetime scales, although the same trade-off principle applies across the entire continuum of time scales. For instance, the pond snail *Lymnaea stagnalis* maintains a particular concentration of sodium ions in its hæmolymph, which, under conditions of low sodium concentrations in the external medium, requires uptake of external sodium ions by active mechanisms in the integument to make up for passive and urinary losses [93]. The fitness benefits of maintaining a stable ionic environment in the fluid that bathes the tissues are traded off against the opportunity costs incurred because energy is invested in transporting ions against the overall medium/hæmolymph gradient, and molecular building blocks are invested in synthesising the molecular machinery that mediates this transport.

As the hæmolymph sodium ion concentration falls, the snail's neuro-endocrine system secretes Sodium-Influx Stimulating peptide (SIS) [93]. The resulting elevated levels of this peptide hormone in the hæmolymph represent the *drive* to maintain a (nearly) constant hæmolymph concentration — the hormone effectively *urges* the integument to take up sodium.

The word *hormone* derives from ὁρμῶν, the present participle of ὁρμᾶν 'to set in motion,' 'to urge,' which in turn goes back to ὁρμὴ, 'urge.' For the Stoics, this ὁρμὴ could be aroused by an external stimulus or internal stirrings of the body. Stoics writing in Latin had various terms for ὁρμὴ, which gave rise to our *impulse*, *impetus*, and *appetite*, this last one from *adpĕtītĭo*, 'that which is sought or desired.'[†] With a little goodwill and hindsight, these Stoic teachings could be construed as the beginnings of neuro-endocrinology.

[†] Stoic philosophers valued the controlled channeling of ὁρμὴ by exerting judgement and self-discipline, which lend or withhold assent (συγκατάθεσή) to these drives [83].

For present-day scientists, hormones are endocrine factors, humoral signalling molecules that are distributed throughout the body via the blood or hæmolymph compartment, as opposed to paracrine, autocrine, and juxtacrine factors, which are confined to the immediate tissue environment [16]. All such factors are called *first messengers* to distinguish them from *second messengers*, signalling molecules that trigger intracellular signalling cascades. Synaptic communication by means of neurotransmitters can be regarded as a special case of paracrine/autocrine signalling [16]. First messengers usually evoke cellular responses by engaging receptors situated at the exterior side of the membrane of the target cell, but they may also engage intracellular receptors, in which case signalling is intracrine [16].

Hormones appear to play a central role as regulators‡ of life history trade-offs: for instance, testosterone mediates the trade-off between reproduction and immunity in the bird *Junco hyemalis* [43]; follicle-stimulating hormone (FSH) mediates the trade-off between egg number and egg size in the lizard *Uta stansburiana* [80]; and juvenile hormone (JH) mediates the trade-off between flight capability and reproduction in the insect *Gryllus firmus* [101]. The choice between investments in somatic mass and endurance on the one hand and direct investment in offspring on the other is among the most consequential life history trade-offs: investment in the former can, with suitable discounting, be leveraged into more of the latter, later in life [12]. It is in this juvenile/larva→adult transition (or more generally, possibly iterated non-reproductive↔reproductive phase transitions) where pleiotropy manifests itself most dramatically, since the different life stages correspond to physiological, anatomical, and biochemical differences that can be profound.

The Endopterygota (or holometabolic insects) constitute an example *par excellence*, with individuals undergoing a radical metamorphosis through a pupal stage in which the entire body mass is reorganised. Accordingly, the pleiotropic suite of JH in the holometabolic insect *Drosophila* includes numerous targets involving virtually all aspects of the phenotype, such as metamorphosis, development and differentiation

‡The use of terms of art such as *regulation*, *control*, *modulation*, *mediator*, etc. does not seem to be well-regimented in the life sciences, even among specialists. For most authors, these terms are virtually interchangeable, perhaps up to some differences in emphasis or nuance — the occasional pedants have a bee in their bonnet about *regulation* versus *modulation*, or philosophical scruples proscribing *control*. In the chapters that follow, we will use *regulation* in the technical sense of feedback-based corrective intervention in system behaviour [64] and mostly reserve *control* for the sense it has in Optimal Control Theory [69].

of adult body structures, reproductive organs, pheromone production, locomotor and courtship behaviour, division of labour in social insects, brain structure and function [25].

The radical realignment of 'purpose' at the whole-organism level — the dramatic shift in key of the organismal harmony — correlates with an extensive suite of JH targets. It can be seen that 'purpose' here begins to acquire a more respectable sense by virtue of the evolutionary background. We will return to this insight several times, leaving the idea fairly intuitive for the moment and instilling it with more rigour in what follows.

From an evolutionary point of view, trade-offs represent balancing points between evolutionary pressures. At the level of neuro-endocrine regulation, the trade-off is a balance of opposing drives, negotiated by hormones and other first messengers. Combining these two perspectives, we perceive first messengers to be *tokens* of the evolutionary pressures (i.e. fitness differentials) that govern such trade-offs.

1.2 A THEORY OF REGULATORY INFRASTRUCTURE

Summarising the argument of the previous section, the central tenet of the emerging field of evolutionary (neuro)endocrinology can be stated as follows:

> First messengers in biological regulatory systems are tokens of evolutionary pressures that govern (life-history) trade-offs.

Connecting messengers and evolutionary pressures in this way may at first seem far-fetched, given that the former act over time scales several orders of magnitude shorter than the typical life span of the organism, whereas the latter act over time scales several orders of magnitude longer. It is also not clear whether the above statement should be understood to be a law or perhaps just a guiding principle. Our task, methodologically speaking, is to strip the concept of 'token' of any metaphysical or mystic connotations, and to this end we shall break the thesis down into a series of posits. For the time being, we discuss matters as informally as we can; the chapters that follow will make the intuitive ideas more precise.

> 1 Control in a biological system tends to drive the system from a given physiological state with an urgency that is proportional to the maladaptiveness of that state.

As stated, this is a rather sweeping statement that is susceptible to criticisms ranging from incoherent to tautological. We certainly have to give precise meanings to terms such as 'urgency' and 'maladaptiveness' and it is one of the aims of the following chapters to make good on this promise. One could also quibble with proportionality — any monotone increasing relationship would do as well. More importantly, the intrinsic cost of regulation is yet to be discounted. Whilst these are important caveats, there does seem to be an obvious underlying truth that, if regulation arises in the course of evolution at all, we should expect, all else being equal, behaviour that tends to steer the organism *away from* states (behaviours, etc.) that decrease its fitness rather than *towards* such states.

The upshot from posit 1 is that natural selection will mould the dynamics of the regulatory system so as to behave as if subject to a variational principle, whose Lagrangian bears a direct relationship to the fitness landscape (more specifically, a marginal fitness projection [12]). A more careful discussion is quite delicate and will not be taken up until Chapter 7.

If we accept posit 1 and grant that biological control is based on 'regulatory drives' that represent fitness gradients, the question arises of how these generalised forces are exerted on the biological system. This is a question about *regulatory infrastructure*. The physiological quantities that we would generally regard as being subject to regulation — for instance, concentrations of electrolytes, nutrients, pigments, enzymes, and the like, as well as surface densities of receptors, ion channels or transporters, temperatures, masses of tissues, compositions of tissues in terms of differentiated subtypes — are being controlled not directly, but through the modulation of rates, fluxes, genes being 'open' and so on. Hormonal pleiotropy means that a given regulatory drive may act through several such actuators, and that each controlled variable may be affected by several actuators as well as by other controlled variables. A mathematical dynamicist would summarise this as follows:

> 2 The connectivity structure of the regulated physiological network is represented by the (sign structure of) the Jacobian matrix of the physiological dynamics.

We have already seen hints of a distinction to be drawn between 'controlled' variables on the one hand and 'actuator' variables on the other; an example is shown in Fig. 1.1. The theory developed in this monograph will frequently draw on this distinction, but it should be emphasised from the outset that the distinction is predominantly a pragmatic one, and that, depending on the problem at hand, interpreted variables can be moved

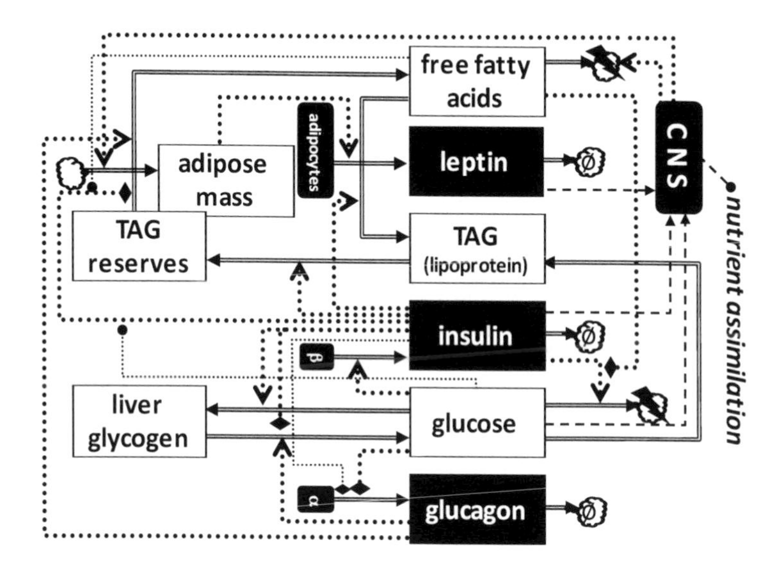

Figure 1.1 **Control of energy metabolism in mammals.** Highly simplified diagram showing a putative division into a 'physiological' component and a 'regulatory' component, the former represented by white boxes with black lettering, the latter by black boxes with white lettering. Fluxes and transformations are depicted as double-barrelled arrows; (neuro)endocrine signalling as dotted arrows. Information flows (sensory monitoring) are represented as dashed arrows. Sources and sinks are drawn as clouds. **TAG**: triacylglyceride; **CNS**: central nervous system; **α**: pancreatic alpha cells; **β**: pancreatic beta cells.

from one category to the other. A heuristic principle that sets the 'controlled' or 'physiological' ones apart is the following:

> 3 Physiological dynamics is determined almost entirely by fundamental physico-chemical principles.

The fundamental constraints imposed by, for example, conservation principles, scaling laws, the physics of transport phenomena, tend to go a long way toward determining the dynamical behaviour of the controlled component of the organismal system, a component that for want of a better word we shall call the 'physiological' component — a slight misnomer, since the actuator variables, and the state variables of the control systems that govern them, are every bit as physiological.

Posit 3 warns us against too cavalier an attitude toward fundamental constraints, when we set about formulating a mathematical model of an organismal system. It is unwise to dismiss such constraints as a minor annoyance to be verified perhaps only as an afterthought. For instance, it may seem obvious (as well as convenient) to treat foodstuffs as sources of calories and represent the organismal energy budget in a single caloric currency. However, as an unintended (and unnoticed) side effect, the theory may implicitly be assuming that nitrogen atoms can be transmuted into carbon atoms by normal biochemical processes. The wise modeller who departs from basic constraints will find that a substantial part of the mathematical modelling effort is already accomplished just by stating the constraints in a precise and coherent manner.

What posits 2 and 3 tell us, when taken together, is that we may be able to work out something like the 'control logic' of a biological system by first taking stock of thermodynamical constraints, biochemical/stoichiometrical constraints, conservation principles (for instance, atomic nuclei are not destroyed or created in [bio]chemical reactions), transport laws (e.g., diffusion laws and scaling effects that arise when surface area-mediated fluxes interact with mixing in three-dimensional volumes) then exploit these constraints as a scaffolding for the dynamics of the physiological component of the system, and finally, deduce from the sign structure of this dynamics how 'regulatory drives' or 'imperatives' would translate into stimulation or inhibition of potential targets. If we cannot liberate endocrinology students from their obligation to memorise pleiotropic suites by heart, then perhaps we can help such suites by showing how to derive them from first principles.

With the physiological component's dynamics *grosso modo* covered by posit 3, the next posit suggests how we may tackle the regulatory component.

> 4 The actuator variables obey gradient dynamics based on a variational principle.

This posit can be appreciated by contrasting it with its diametrical opposite, which is a conventional, mechanistic bottom-up approach that combines the electrophysiology of messenger-secreting cells (neurons as well as endocrine cells) with intracellular metabolism, signalling cascades and genomic and transcriptome dynamics. All such mechanistic fine detail is certainly fascinating and worth modelling in detail, but the *curse of dimensionality* looms large here: detailed modelling of (neuro)endocrine circuitry readily leads to models where the number of dynamic degrees of freedom of the regulatory part dwarfs the physiological part by several orders of magnitude.

To be sure, these problems can be surmounted, and there is nothing inherently or conceptually wrong with 'big' models. The thinking behind posit 4 is rather that it is useful to have a minimalistic default model, which can be scaled up to accommodate any desired level of complexity or detail. From a mathematical modelling perspective, posit 4 is primarily a pragmatic one. It suggests a systematic model reduction approach to the regulatory component, closely related to well-established and familiar ideas such as *lumping* and *coarse graining*.

Unfortunately, posits 1 and 4, taken together, may strike one as outrageous philosophy. Purposefulness, or the appearance of it, is a perennially contentious issue in biology [30, 38, 61]. Whenever we discern such apparent purposefulness in biological systems, we may choose to view this as an expression of τέλος, a final end point or inherent aim which is already intrinsic in the nature of a thing. Classical antiquity adhered to a *natural teleology* which endowed the laws of nature itself with such intrinsic purposefulness [3], which is perhaps best viewed as a natural, unforced unfolding rather than a conscious striving. Intrinsic τέλος is cut of the same cloth as the Stoics' impetus or ὁρμή, and thus we should not be too surprised that hormones have landed us in the hot water of teleology.

Moreover, the preponderance of variational principles in contemporary theoretical physics may encourage the belief that teleological thinking could be central to the natural sciences. Aficionados, who count among their number a luminary no less than Max Planck [100], should heed the nominalist caution that mathematical formalism has no privileged position as a putative wellspring of metaphysical truth.

In contrast to intrinsic τέλος, *extrinsic purposefulness* resides in the use and design that is imposed on objects or fellow creatures by beings capable of planning. In the context of monotheistic faith, such as the Abrahamic religions, intrinsic and extrinsic purposes have a tendency to come to the same thing, as they both flow from the sheer will of a personal, omnipotent, and benevolent supreme being. This is also true for other belief systems, insofar as power/ownership, volition, and personal consciousness are ascribed to super-natural phenomena. The fundamental distinction between the apparent design of a biological control loop and the engineering design of a throttle governor then becomes moot, because either one is taken to imply the presence of a conscious mind that has arranged apparatus (be it biological or brass) based on 'reasoning backwards' from desired final states.

On the other hand, the doctrine of natural teleology maintains a strict distinction between intrinsic and extrinsic purposefulness; it is still occasionally put forward as a tenable explanation for (apparent) purposefulness in the life sciences [61]. This strict division does not mean that natural teleology is necessarily atheistic or agnostic, although it most readily associates with such stances.[††]

To appreciate that evolutionary dynamics is quite sufficient to account for the connections made by posits 1 and 4, let us suppose that gradient-driven dynamics provide a reasonably accurate coarse-grained model of the workings of the regulatory component, as corroborated by several case studies in ref. 14. If we allow, in addition, that the potential function ('Lagrangian,' see Chapter 3) of this gradient dynamics is subject to genetic variation, then we could readily admit that the extent of agreement between the fitness landscape and this potential function will correlate positively with the degree to which life histories (in the sense of fully realised developmental paths) culminate in higher reproductive output — and such correlation suffices to complete the evolutionary argument [12].

We may be satisfied that we need not invoke teleological metaphysics, provided that we can establish that the potential function which drives the gradient dynamics of the regulatory component is an 'evolvable' function. By this we mean that it can be tweaked by natural genetic variation to shift maxima, minima, and saddle points, or to steepen up certain portions or flatten others, and so on.

But this function is eminently evolvable. If there is any embarrassment here at all, it is an overwhelming embarrassment of riches. The plethora of molecular and cellular entities that constitute control loops represents a equally abundant cornucopia of variation: sensitivities of sensor molecules, permeabilities and gating characteristics of ion channels, affinities of messenger molecules and their target receptors, production and turn-over rates of these messengers, numbers of specialised cell populations involved. The list is endless and so are the points where natural selection may gain purchase.

The situation is well-captured by the celebrated visual metaphor due to Waddington (originally proposed in the context of developmental pathways, but equally if not more apt here), which is depicted in Fig. 1.2. If we take the controlled physiological state to be two-dimensional (for

[††]Natural teleology can be conceived as compatible with deism or suitably adapted forms of theism, such as pantheism, or even monotheism, provided that the divinity is primarily conceived of as first efficient cause. The rejection of natural teleology, which shall be the mainstream position assumed in this monograph, should not be seen as a disparagement of any side in religious debates.

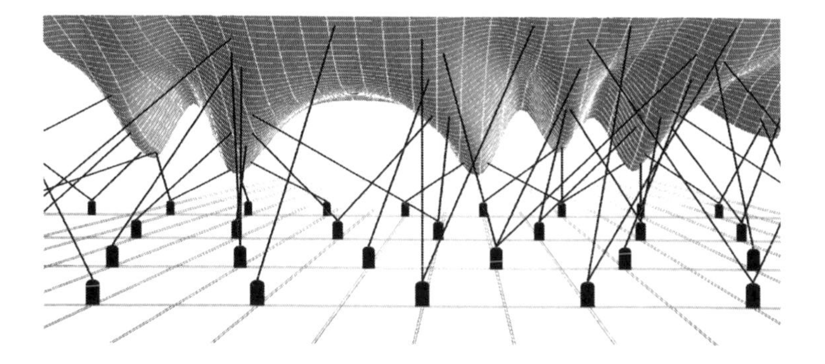

Figure 1.2 **The epigenetic model of the regulatory potential function.** See text for further explanation.

the sake of picturesqueness), the associated potential function can be represented as a surface plot; we imagine this surface as a rubber sheet, seen from below in Fig. 1.2, tugged by guy-ropes connected to pegs in the ground. The pegs represent genes, and 'the strings leading from them the chemical tendencies the genes produce' [86]. In this visual metaphor, genetic variation alters parameters such as the length of the strings, the tension they support, and the points where they attach to the rubber sheet.

1.3 SCOPE & OVERVIEW

Every organism has at its disposal a finite supply of time, energy, and molecular building blocks. It stands to reason that life-time reproductive success critically depends on how the organism allocates these scarce resources to various anatomical structures, tissues, behaviours, eggs, parental care, and so on. The manner in which the organism takes decisions regarding differential allocation has a genetic component: the apparatus that mediates these choices (neural and glandular tissue, sensory organs) is genetically encoded, as are perhaps the tendencies and biases that inhere these choices. Thus, provided there is some genetic variation, it is meaningful to say that natural selection will favour those 'allocation types' that navigate life history trade-offs that maximise inclusive fitness [12].

A major problem in evolutionary biology is evo-devo indeterminacy at the phenotypic level. In terms of Waddington's visual metaphor, the guy ropes (genetic-developmental underpinnings) may be arranged in myriad different ways to achieve the same surface topography of the landscape [72]. For example, such indeterminacy manifests itself in the form of convergent evolution [89]. Natural selection responds only to the results, not to how these are attained: it is — to a substantial extent — blind to 'inner workings.'

This problem is particularly pressing in the case of regulatory systems [72, 89]. During the development of the individual organism, the decision apparatus represents an investment in its own right. Consequently, there must be a selective advantage to having the same information processing capability with fewer molecules, cells, and so on. Up to this performance standard, the 'wiring' of the apparatus is evolutionarily neutral.

It is just this indeterminacy that bestows empirical content on the proposition that hormones betoken evolutionary pressures: the thesis put forward is that the fitness landscape is actually mirrored by the regulatory architecture that underpins the organism's differential allocation choices. The central objective of this monograph is to give a precise meaning to this proposition. In Section 1.2 we set out a program of work (posits 1–4), the core of which will be developed in Chapters 2 and 3.

Subsequent chapters will elaborate on the central thesis. For instance, posits 2 and 3 together imply that pleiotropic suites should be derivable from physiological constraints and trade-offs; one possible approach to this problem is explored in Chapter 4.

The physiological component corresponds roughly to the 'controlled' part of the organismal system, whereas the regulatory component comprises the 'controller' [102]. However, the boundary between the physiological and the regulatory components is a pragmatic one; we are not insisting on any fundamental distinction that sets the two apart. As argued in the previous section, the motivation for drawing this distinction is primarily that of keeping the curse of dimensionality at bay, by offering a modelling approach that is minimalistic with respect to the number of dynamic degrees of freedom. The state space dimension of physiological component can be as low or high as prescribed by the modeller's needs, available data, and so on.

The use of gradient-driven dynamics for the regulatory component can be justified, however tentatively, as a form of coarse-graining. In Chapter 5, we take this one step further and consider how the state of the regulatory component can be collapsed entirely. The state space of the physiological-plus-regulatory components can be viewed as the Cartesian product of a subspace spanned by the 'physiological' state variables and another one spanned by the 'actuator variables.'

5 The physiological component can be modelled as a differential inclusion, which arises as a limiting case when the actuator variables are eliminated via a time-scale argument.

The time-scale argument is not always available, as the fastest modes of the physiological component may be faster than the slowest modes of the regulatory component. However, whenever there is a clear separation between these modes, the collapse of the state space onto just the physiological subspace is possible, and the natural framework for the resulting dynamics is provided by the theory of differential inclusions.

The broader interest in posit 5 is that one often encounters mathematical models of biological systems whose behaviour is (co-)governed by the internal control systems that are not represented in the model (presupposed, to be sure, but not explicitly expressed in the model's mathematics). Examples abound and include models of growing (or waxing) cell populations, e.g., in developmental biology, infection and immunity, cancer, as well as models of population dynamics at the ecological level [11].

Virtually any mathematical model of a biological process must in some sense represent a reduction, a projection onto a manageable number of dynamic degrees of freedom. This truism does not invalidate such models. But we ought to ask how this reduction is possible at all, and also how we can make hidden assumptions about control systems explicit, and systematically go back and forth between models that make (some subset of) details explicit and 'reduced' models.

To demonstrate the collapse onto a differential inclusion, we take up a case study in Chapter 6, which concerns the nutrient/energy balance of a generic mammal at the organismal level, with a view to exploring both the strengths and the shortcomings of the approach suggested by posit 5. This case study illustrates how fundamental principles (specifically, stoichiometric conservation and surface area / volume scaling laws) suffice to define the dynamics of the system, up to remaining degrees of freedom, where a variational principle encoding homeostatic drives defines the qualitative dynamics (i.e., geometry of flow in the phase space).

The case study of Chapter 6 also brings out an emergent qualitative effect:

6 Interlocking control loops can give rise to frustrated physiological states.

Control loops *interlock* when they share actuator variables, and states are *frustrated* when controlled variables are prevented from attaining their respective physiological optima. Frustration would not occur if each controlled variable had its 'private' actuators, that is, if there were no pleiotropy. The control loops would then function independently and would each be separately susceptible to the elementary analysis reviewed in Chapter 2.

Admittedly, it seems obvious that such an effect might occur whenever control loops become entangled through shared actuators. Nonetheless, the analysis detailed in Chapter 6 suggests that it may be possible to interpret and understand some of the complexity of regulation at the organismal level as arising from interlocking control loops; this perspective deserves to be more fully explored.

CHAPTER 2

The nature of homeostasis

Homeostasis refers to the tendency to maintain a constant or near-constant internal environment for the biochemical and physiological processes of life [102]. Familiar examples at the cellular level include the cytosolic concentrations of chemical species that affect the rates of metabolism (e.g. $[H_3O^+]$, [ATP], the intermediates of the Krebs† cycle) and at the organismal level physico-chemical parameters of interstitial fluids and blood, such as temperature, pressure, oxygenation, nutrient concentrations [78]. By maintaining such variables at constant or near-constant values, the body creates a stable and predictable *milieu intérieur* providing steady operating conditions for the processes of life, even when the organism's ambient environment, the *milieu extérieur*, undergoes marked fluctuations or is severely perturbed.

At the heart of homeostasis lies the principle of *feedback control*, also known as *closed loop* control. Life scientists perhaps shy away from the term 'control' to avoid unintended teleological imputations and may prefer to speak of *regulation*; however, the hallmark of regulation is the use of feedback to attain what engineers unabashedly call 'desired behaviour' [64]. In a biological context, 'desirable' has to be understood as a shorthand for 'conferring adaptive value,' as we will argue in Section 2.1. Moreover, the biological context requires a slightly different mathematical formalism than the classic engineering context, because coarse-graining and time-scale arguments are essential

†By *Krebs cycle* we shall invariably mean the tricarboxylic acid cycle, also known as the citric acid cycle — not the urea cycle which has an equal claim to the 'Krebs' appellation [77].

for tractable models of complex biological systems [11]. The basic notions and notations used in the chapters that follow are introduced in Section 2.2.

2.1 THE ORIGINS OF HOMEOSTASIS

Information science in its broadest meaning presents itself as the medium of choice for the study of biological regulation. Control and homeostasis find a natural fit in what used to be called *cybernetics*, the theory and craft of process control (from κυβερνητική, the art of the maritime pilot). However, there are differences in outlook and parlance which engender unnecessary confusion and debate. For instance, control engineers call the value that is maintained the 'reference value,' or 'desired value,' or 'set-point' [39], terms which may worry the biologist, since the system operates without conscious desires or goal-orientation [38].[‡] In the present section, we examine the biological origins and nature of homeostasis, in an attempt to dispel such confusion.

Let us consider an enzyme: we know that its catalytic efficiency is a function of temperature, with an 'optimum' defined by the enzyme's intrinsic physico-chemical properties [67]; this optimum is certainly without teleological implications, and can simply be thought of as an extremum (maximum) in the mathematical sense of the word. At the same time, we readily concede that it is subject to evolutionary change, as mutations in the enzyme's amino acid sequence will shift the optimum temperature, and the evolutionary pressures will tend to make the optimum agree with the prevailing ambient temperature. The latter may simply be the environmental temperature of the organism. Fluctuations in this temperature may cause periods of reduced catalytic efficiency (and hence reduced metabolic rate), which may either be endured or counteracted in various ways. One such way is the expression of several variants of the enzyme, *isoforms*, each with its own optimum [36]; thus the overlapping bell-shaped curves ensure quasi-independence on temperature over a wide range. However, this solution requires the investment of molecular building blocks into several isoforms, or else regulatory apparatus to adjust gene expression.

[‡]These misgivings include (i) that it is naïve to countenance optimal control or control engineering ideas in biology (avoidance of shibboleths is not in itself a cure for naïveté); (ii) that 'set-point' implies perfect regulation over some range of inputs (it does not); and (iii) that set-points are pre-set targets whereas in biology they are emergent network properties (any model in biology that is not wholly in terms of atoms and molecules has 'emergent' properties for its parameters, so why single out set-points?).

If the enzyme could be exposed to a temperature that fluctuates to a lesser extent than the environmental temperature, it could serve its role as a cog in the larger metabolic and physiological process more efficiently and more reliably. This can be achieved if the cell in which the enzyme works is part of a multicellular assemblage that creates its own *milieu intérieur* which is more stable and predictable than the *milieu extérieur.*

The gain in efficiency is due to a diminishment of the need to effect adaptations locally, at the level of the single cell. Acellular ('single-celled') organisms are capable of adaptation to harsh, fluctuating environments, but they have to serve the regulatory operations of sensing and adjusting gene expression locally; similarly, they have to absorb changes in nutrient availability by creating storing excess that is assimilated in times of plenty in the form of cellular inclusions [50]. Being part of an aggregate that creates its proper *milieu intérieur* allows cells to shed many of these functions whilst excelling at some (specialisation) which calls for different gene expression patterns with corresponding differences in cellular biochemistry and morphology (differentiation). Various species that are traditionally classified as single-celled are in fact somewhere along this transition to aggregates of specialised cells [48].

Local autonomy is thus traded for greater reproductive success (inasmuch as the aggregate as a whole is better at extracting energy and matter from the environment). The cost of contributing to the global aggregate is far outweighed by the benefits of being part of it. Ultimately, that trans-generational genetic transmission becomes a specialised function itself. This argument for cooperation within an aggregate, and contribution to a shared nourishing and stable environment, not only applies to specialised cells within a multicellular organism, but also at lower levels of organisation (organelles within a eukaryotic cell) and higher levels (individuals within a colony or hive).

There remains a soupçon of circular reasoning in this account: homeostasis, in the sense of provision of a reliably stable *milieu intérieur*, rests on the efficacy of the specialised tissues, which in turn rely on homeostasis. It is true that, in the fully evolved system, we encounter a state of mutual dependence in which no component can work without the other. However, there already is benefit (and hence adaptive value) in *some* degree of alleviation of ambient variability and unpredictability, however imperfect. Thus, differentiation and specialisation, e.g. of light-capture, storage, and reproductive roles [45] may come first, and then as the multicellular aggregate attains a certain size, a pool for sharing nutrients, gases and minerals, waste products, and heat naturally arises. This sets the stage for the gradual relegation of local protective mechanisms to aggregate-level regulation of the physico-chemical characteristics of this pool, which is typically an interstitial liquid that bathes the cells and tissues, and that may become further compartmentalised into hæmolymph, blood, lymph, CSF, and so on.

What we have coyly been calling the 'aggregate' is of course none other than the individual organism. Individuals can be *unitary* or *modular*, and with certain modes of reproduction, the distinction between a modular organism and a colony of unitary ones may become blurred [4]. However, such distinctions do not materially affect the present discussion — except perhaps with respect to the question of to what extent the homeostatic regulatory system can be centralised or must remain distributed — and in fact the evolutionary story of a gradual relegation of local autonomy to aggregate-level control should lead us to expect a blurred boundary between colony and individual.

As soon as homeostatic regulation is capable of achieving a near-constant value for one of the physico-chemical characteristics of the interstitial pool, the question arises as to whether this value is in any meaningful sense an 'optimum.' As the foregoing argument suggests, homeostasis arises as a *boot-strap*, a process of co-evolution. To return to the example of temperature, if the system becomes strongly homeothermic with a reigning internal temperature of T^* Kelvin, then what matters for adaptiveness is that the optima of the enzymes in the organism are at or close to T^* as well (the 'optimum' of an individual enzyme has a clear-cut interpretation, as we saw).

The requirement of *mutual co-adaptation* is more central than the value of T^* as such (several eco-physiological requirements do impinge on the evolutionarily most adaptive value of T^*). Matters become even more complicated if we factor in evolutionary contingency. For instance, T^* may be well above the range of ambient temperatures experienced by the organism, affecting its heat balance and perhaps necessitating secondary adaptation such as thick fur or blubber [78]; but the direct ancestor of the species may have been exposed to a *milieu extérieur* with temperatures close to T^* during the active part of the sleep/wake cycle. If in the present species T^* is well in excess of the ambient temperatures, we should perhaps expect T^* to evolve downwards. However, this requires co-evolution of almost the entire proteome, which would tend to bring this downward adjustment almost to a halt.

These examples suggest an obverse side of the 'homeostatic boot-strap': the system tends to be 'locked in.' Another possible example of this is the fact that in many animals, $[Na^+]/[K^+] \gtrsim 10$ for blood or hæmolymph, much like sea water, but in marked contrast to cytosol, where $[Na^+]/[K^+] \approx 0.1$; this suggests that the interstitial liquid still mimics the sea, at least in some of the respects that most affect resting membrane potential and excitability [88].

Thus, the 'target value' or *apparent set-point*, for instance, the value T^* in the foregoing discussion, need not define an optimum in any absolute or objective sense. Nevertheless, departing from the observation that the peripheral physiology and biochemistry is to some degree 'locked in,' we may take the properties of the latter to direct the evolution of the regulatory system.

Apart from the spectre of teleology, which we hope to have dispelled, the set-point concept also raises concerns regarding its biological reification. In engineering applications of feedback control loops, the set-point can be realised as a discrete, delineable physical entity, such as a potentiometer that is set to provide the reference value. In biological systems, one looks in vain for a component, be it a molecule, organelle, dendrite, or the like, that can be isolated and designated as the physical realisation of the (apparent) set-point of the system.[††] If a set-point is not an identifiable *thing* within a biological system, it would seem to be unevolvable *a fortiori*. This oft-heard criticism of the set-point concept is misguided. A counterexample is given by the following example, a highly simplified model of a gene expression control loop.

Example 1: enzyme gene expression. Consider the following system of ordinary differential equations:

$$\dot{x}_1 = u - x_2\left(1 + x_1^{-1}\right)^{-1} \tag{2.1}$$

$$\dot{x}_2 = \alpha x_3 - \mu x_2 \tag{2.2}$$

$$\dot{x}_3 = \beta x_1 x_2 - \lambda x_3 \tag{2.3}$$

where α, β, λ, and μ are positive parameters, x_1 represents a metabolite, x_2 is an enzyme for which this metabolite is a substrate, and x_3 is the mRNA encoding the enzyme (x_1, x_2, and x_3 are concentrations in the cytosol of cell); $u(t) > 0$ represents the net flux due to other processes that produce or consume the metabolite. When $u(t) \equiv u^* > 0$, x_1 will tend toward the value $x^* = \mu\lambda(\alpha\beta)^{-1}$ irrespective of $\overline{u}$, and when $u(t)$ fluctuates, $x_1(t)$ fluctuates about x^* with amplitudes and frequencies that depend on $u(t)$. In intuitive terms, when μ and λ are 'large' enough relative to the peaks in rate of change of $u(t)$, $x_1(t)$ will tend to track x^* closely. ❖

[††]Nevertheless, it has been claimed that transfer of control concepts to biological systems presupposes or requires that a modular molecular or cellular equivalent of a set-point, mimicking a 'pot-meter' or similar, be postulated [19].

Any biologist acquainted with the regulation of gene expression no doubt itches to add details and realism to this very basic example. Despite its simplicity, the example does show us that an *apparent set-point*, here x^*, can arise as a *compound parameter* of 'mechanistic' parameters (in this case $\mu\lambda(\alpha\beta)^{-1}$). By 'mechanistic' parameters we mean quantities such as rates, affinities, permeabilities, and so on, which are unproblematic in the sense that we can readily conceive how mutations in the corresponding proteinaceous machinery lead to small adjustments of the values of these parameters. Thus we see how (apparent) set-points are *distributed characteristics* of the system. As such, set-points are true traits, i.e., susceptible to evolution by natural selection.

2.2 THE ELEMENTARY HOMEOSTATIC LOOP

The use of feedback information (i.e, information about the state of the regulated system) is the defining characteristic of regulation [64]. In the chapters that follow, we shall be interested in what happens when regulatory circuits become entangled and wholly or partially conflicted. To prepare for this, we here introduce notation that accommodates both simple loops and interlocking loops.

In a closed-loop or feedback configuration, there is an information flow from controlled system to controller and back (Fig. 2.1). the controlled variable is monitored by a dedicated sensory apparatus, which conveys an afferent signal to groups of cells that process this information; these in turn generate efferent signals that induce changes in one or more physiological processes, having the net effect of counteracting the disturbance and restoring the controlled variable to the homeostatic value [102]. The information-processing tissue may be neural, endocrine, or neuro-endocrine [90].

Control systems generally act by directly adjusting certain quantities (called *actuators*) in such a way that the homeostasis-relevant quantities remain confined to a particular range of values.

Example 2: temperature control. Temperature homeostasis may be modelled in a highly simplified form by the following ordinary differential equation:

$$\dot{x} = u - x + z \tag{2.4}$$

where x is temperature of the body core, $\dot{x}$ its rate of change, u the ambient temperature, and z heat production, and all variables have been

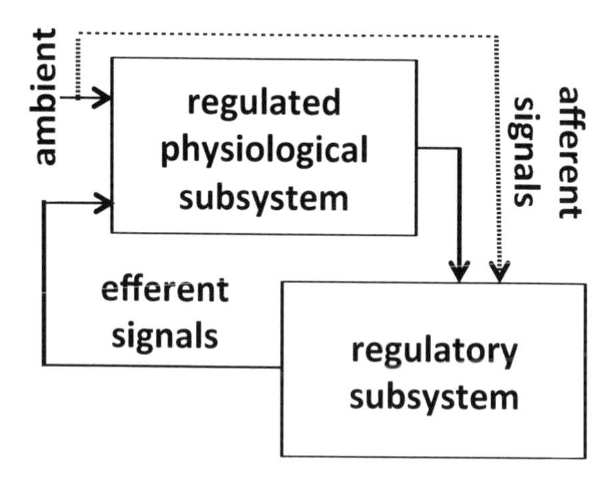

Figure 2.1 **Elementary control loop.** Arrows indicate the travel of information between the subsystems. Afferent signals carry information towards the regulatory subsystem; the dashed line indicates information about the ambient input or disturbance being relayed directly to the regulatory subsystem (feedforward) whereas the solid line indicates physiological information (feedback). Either or both of these constitute the input to the regulatory subsystem; its output is marked 'efferent' and corresponds to physiological processes (rates, fluxes, etc.) that are adjusted in such a manner that homeostasis is safeguarded. The regulated subsystem receives both the efferent input and the forcing imposed by the ambient.

scaled appropriately. Here, x is a *physiological state variable* and z is a *physiological actuator variable*. We can suppose without loss of generality that the scaling is such that the physiological optimum corresponds to $x = 1$. If the ambient conditions are constant, $u(t) \equiv u^*$, homeostasis may be attained by setting $z^* = 1 - u^*$, provided that the latter value can be realised by the system. Accordingly, we require that z^* is contained in the interval $[z_{\min}, z_{\max}]$. Strict homeostasis will fail unless $z^* \in [z_{\min}, z_{\max}] \Leftrightarrow u^* \in [1 - z_{\max}, 1 - z_{\min}]$; the latter is the *regulatory range*. To understand the behaviour of the model outside of this regulatory range, we suppose that z is set to $z_{\max}$ whenever $u < 1 - z_{\max}$. This yields a steady-state in which body temperature is below the optimum, but minimises the difference between 1 and the steady-state value attained ($x^* = u^* + z_{\max}$). For the same reason, we would expect that z is set to $z_{\min}$ when u^* exceeds the top of the regulatory range ($1 - z_{\min}$). This is a general pattern: outside of the regulatory range, the actuator variable is 'stopped' at an extreme value, and the regulated physiological state variable fails to be held at the homeostatic point. ❖

This example can be generalised to higher dimensions: we let $\boldsymbol{x} \in \mathbb{R}^n \equiv \mathcal{X}$, $\boldsymbol{z} \in \mathbb{R}^m \equiv \mathcal{Z}$, $\boldsymbol{u} \in \mathbb{R}^q \equiv \mathcal{U}$, where n, m, and q are positive integers. The main motivation to introduce this distinction between x-type variables and z-type variables is that the former lend themselves more readily to basic physico-chemical modelling, which yields dynamics $\boldsymbol{f}$ such that

$$\dot{\boldsymbol{x}} = \boldsymbol{f}(\boldsymbol{x}, \boldsymbol{z}, \boldsymbol{u})\,. \tag{2.5}$$

We shall occasionally write $\mathcal{X}\sim$ to mean 'pertaining to a physiological state variable (x-type variable)' and similarly $\mathcal{Z}\sim$.

From a physiological point of view, we regard $\boldsymbol{z} \in \mathcal{Z}$ as determined by (as well as representing) an underlying *regulatory state* $\boldsymbol{x}_{\mathrm{reg}} \in \mathbb{R}^N$. From the point of view of mathematical modelling, our options are to specify the dynamics of $\boldsymbol{x}_{\mathrm{reg}}$ outright, or to 'close' the system, motivated by the consideration $m \ll N$, either by specifying dynamics for $\boldsymbol{z}$ or by specifying $\boldsymbol{z}$ as a function of $\boldsymbol{x}$ and $\boldsymbol{u}$. Thus, the state of the system is decomposed into $\boldsymbol{x}$ and $\boldsymbol{x}_{\mathrm{reg}}$, which are the dynamic states of, respectively, the *regulated physiological subsystem* and the *regulatory subsystem* as depicted in Fig. 2.1). Which part of the state belongs to $\boldsymbol{x}$ and which to $\boldsymbol{x}_{\mathrm{reg}}$ is to some extent arbitrary; in broad terms, $\boldsymbol{x}$ is the part whose dynamics $\boldsymbol{f}$ can be described largely by appealing to physico-chemical principles, general physiology, biochemistry, anatomy, etc., whereas $\boldsymbol{x}_{\mathrm{reg}}$ is the part whose dynamics $\boldsymbol{f}_{\mathrm{reg}}$ involve knowledge of neural and endocrine circuitry, first and second messenger pathways, gene regulatory loops, and so on.

The life scientist analyses the behaviour of actuator variables in terms of the complex interplay between various (macro-) molecular species, cells, tissues, and biochemical/physical transformations [78, 102]. In Example 1, the regulatory state $\boldsymbol{x}_{\mathrm{reg}}$ might be said to consist of the variables $\{x_2, x_3\}$ and thus $N = 2$; however, this number is low because the example is highly simplified, and N would quickly increase in any more detailed treatment of the signalling cascade that leads to the regulation of transcription, the kinetics of mRNA (in various stages of maturity, and in various sub-cellular compartments), and translation. Similarly, in Example 2, the physiology would focus on temperature sensors, heat-generating processes, and the neuronal and endocrine pathways that connect them. Given detailed knowledge of all these entities, an explicit representation of their dynamics could be formulated, which would take on the following schematic form:

$$\dot{\boldsymbol{x}}_{\mathrm{reg}} = \boldsymbol{f}_{\mathrm{reg}}(x, \boldsymbol{x}_{\mathrm{reg}}, u)\,. \tag{2.6}$$

Here $\boldsymbol{x}_{\mathrm{reg}} \in \mathbb{R}^N$ denotes the state of the regulatory system and $\boldsymbol{f}_{\mathrm{reg}}$ the dynamics. The actuator variable z depends on this state:

$$z = \boldsymbol{g}(\boldsymbol{x}_{\mathrm{reg}}) \tag{2.7}$$

where $\boldsymbol{g}$ is a function $\mathbb{R}^N \to \mathbb{R}^m$. The information flow is represented schematically in Fig. 2.1.

The dimension N tends to go up quickly when we strive to accommodate any detailed knowledge we might have concerning the inner workings of body temperature control. On the other hand, we may be able to reduce N, for instance by making simplifying assumptions, or by excluding fast dynamic modes in $\boldsymbol{f}_{\mathrm{reg}}$ using time scale arguments. If we succeed in reducing N all the way down to zero, we arrive at a representation of the following form:

$$z = \overline{\boldsymbol{g}}(\boldsymbol{x}, \boldsymbol{u}) \tag{2.8}$$

for a suitable function $\overline{\boldsymbol{g}} : \mathcal{X} \times \mathcal{U} \to \mathcal{Z}$. This closes the dynamics of the physiological system (cf. Eq. (2.5)):

$$\dot{\boldsymbol{x}} = \boldsymbol{f}(\boldsymbol{x}, z, \boldsymbol{u}) = \boldsymbol{f}(\boldsymbol{x}, \overline{\boldsymbol{g}}(\boldsymbol{x}, \boldsymbol{u}), \boldsymbol{u}) \equiv \widetilde{\boldsymbol{f}}(\boldsymbol{x}, \boldsymbol{u}) \tag{2.9}$$

($\boldsymbol{x}$, z, and $\boldsymbol{u}$ are all understood to be functions of time t).

In the case $n = m = q = 1$, it is possible to reconstruct the function $\overline{\boldsymbol{g}}$ (Fig. 2.2). Let us assume that the system settles on a unique steady state when a constant value u^* is imposed; denote this steady state by (x^*, z^*) and assume that we may observe the steady-state response $x^* = h(u^*)$. We might be able to obtain an expression for h by fitting some suitable empirical formula to observational data.

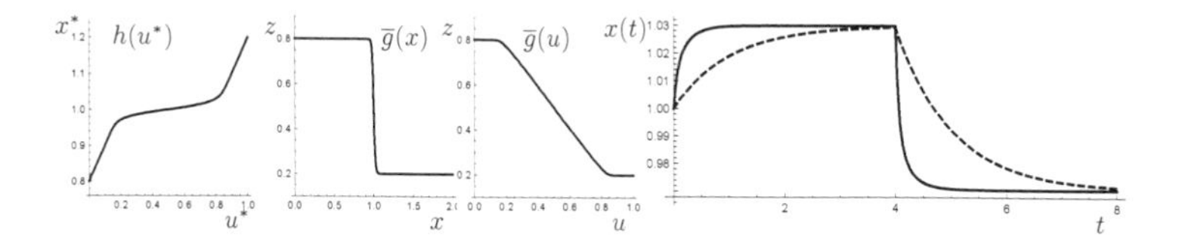

Figure 2.2 **Inferring pure feedback and feedforward controls for scalar loops.** See text for further explanation.

Steady-state is defined by the condition $f(x^*,\overline{g}(x^*,u^*),u^*) = 0$, but in virtue of $N = 0$, this carries over to the transient case, thus:

$$f(x,\overline{g}(x,u),u) = 0\,. \tag{2.10}$$

Since the dynamics f is known, Eq. (2.10) defines $\overline{g}$ implicitly. If we set $x = h(u)$, which is permissible since transient and steady-state relations are interchangeable when $N = 0$, and differentiate with respect to u, we obtain a differential condition describing the function $h(\cdot)$:

$$h' = -(f_u + f_z\overline{g}_u)(f_x + f_z\overline{g}_x)^{-1}\,.$$

If, in addition, we have $\overline{g}_u = \partial\overline{g}/\partial u \equiv 0$, the formula simplifies to $f(x,\overline{g}(x),h^{-1}(x)) = 0$ where h^{-1} is a (local) inverse of the steady-state relationship. On the other hand, if $\overline{g}_x = \partial\overline{g}/\partial x \equiv 0$, the formula simplifies to $f(h(u),\overline{g}(u),u) = 0$.

Applying these ideas to Example 2, we reconstruct the steady-state response $x^* = h(u^*)$ by plotting the two variables against one another and fitting a suitable empirical formula to the data. This might result in a graph somewhat like Fig. 2.2 (left-most panel). In the special case $\partial\overline{g}/\partial u \equiv 0$, we find $\overline{g}(x^*) = x^* - h^{-1}(x^*)$ and hence we have $\overline{g}(x) = x - h^{-1}(x)$ (Fig. 2.2; middle left panel), whereas in the special case $\partial\overline{g}/\partial x \equiv 0$, we find $\overline{g}(u^*) = h(u^*) - u^*$ (Fig. 2.2; middle right panel).

In general, we refer to instances where z depends only on $\boldsymbol{x}$ (not $\boldsymbol{u}$, i.e. $\partial\overline{g}/\partial u \equiv 0$) as *pure feedback* and instances where z depends only on $\boldsymbol{u}$ (not $\boldsymbol{x}$, i.e. $\partial\overline{g}/\partial x \equiv 0$) as *pure feedforward*. The two types of responses are compared in Fig. 2.2 (right-most panel) for the example of Eq. (2.4), in response to step changes in u. Feedback is indicated by the solid line; it can be seen to be faster than the feedforward response (dashed line). The two modes differ in several other respects. For instance, if there is random noise on the afferent signalling line (e.g., $x(t)$ is registered by the regulator as $x(t)+\epsilon(t)$, the latter term being a small white-noise component), the variable z can become quite jittery as the $\epsilon(t)$-term forces z to fluctuate between $z_{\min}$ and $z_{\max}$. This problem becomes more pronounced as the slope of the middle portion of $\overline{g}(x)$, called the *gain*, increases; the problem might be controlled by reducing this slope, but the slope of h will then increase as well, resulting in a greater variation of x over the regulatory range. By contrast, whereas feedforward control can achieve a zero slope for h, it is susceptible to systematic discrepancies between the sensors that register u and the regulatory subsystem that sets z, which are not being corrected by feeding back information regarding the state of the controlled system.

Example 3: hæmolymph sodium concentration in an aquatic invertebrate. The sodium concentration in the blood of an aquatic invertebrate can be represented by the following equation

$$\dot{x} = \alpha(u - x) + \beta u - (\beta + \gamma(x - u))(1 - z)x \tag{2.11}$$

where u is the sodium concentration in the ambient water α, β, and γ are positive constants, and $z \in [0, 1]$ is the fraction of the primarily ultra-filtrated sodium that is reabsorbed in the kidney [95]. The term $\alpha(u - x)$ represents net uptake of sodium due to diffusion of sodium ions through the skin [91]; the term βu represents sodium uptake concomitant with active water uptake (drinking [92]); and the term $\gamma(x - u)$ represents passive water uptake through the skin (osmotic swelling [1]).

Exposing animals to a constant ambient salinity u^* and, after having allowed them to acclimatise to this medium, determining their blood salinity x^*, and repeating this for a range of ambient salinities, yield the data from which we may derive a graphical (empirical) reconstruction of the steady-state relationship $x^* = h(u^*)$. If we assume pure feedback, we are then able to determine $\overline{g}$ in the following form:

$$\overline{g}(x) = \frac{\left(x - h^{-1}(x)\right)(\alpha + \beta + \gamma x)}{x(\beta + \gamma(x - h^{-1}(x)))} \tag{2.12}$$

whereas for pure feedforward we would have

$$\overline{g}(u) = \frac{(h(u) - u)(\alpha + \beta + \gamma h(u))}{h(u)(\beta + \gamma(h(u) - u))} \, . \tag{2.13}$$

The forms of Eqs. (2.12) and (2.13) might seem to presuppose that the system controlling renal sodium reabsorption somehow 'knows' what the steady-state response is supposed to be, as well as the values of the parameters, but this is somewhat misleading. The control system (under the $N = 0$ or 'static' assumption) is simply characterised by a function $\overline{g}$ that reflects the control system's internal properties. The human observer is able to discover this function by adducing the additional information: h is obtained by setting $z = \overline{g}$ and $\dot{x} = 0$ in Eq. (2.11). We are effectively following this logic backwards when our aim is to reconstruct $\overline{g}$. ❖

Reduction to $N = 0$ may not always be possible or biologically realistic. However, reduction down to $N = m$ may be possible, and in this case various other interesting properties emerge.

Example 4: enzyme gene expression (continued). Define an actuator variable via the following ordinary differential equation:

$$\dot{z} = \eta z(x - \xi). \tag{2.14}$$

To link this to Example 1, set $\eta = \beta/\lambda$, $\xi = \mu\lambda(\alpha\beta)^{-1}$ and make the identification $x_1 \leftrightarrow x$. For λ sufficiently large, we approach the identification $x_2 \leftrightarrow z$. ❖

In this example, the steady-state value is ξ, independent of u; i.e., $h' \equiv 0$ in terms of the steady-state response. This is sometimes referred to as 'infinite gain,' but this term is somewhat misleading since *gain* more properly refers to the slope of the function $\overline{g}(x)$, which does not arise in this context. Indeed, Eq. (2.14) exemplifies *integrating* control, whereas *gain* is properly speaking a property of a *proportional control system* [39].

2.3 ELEMENTARY LOOPS ACTING IN CONCERT

The previous chapter suggested that organismal harmony arises from the combined operation of multiple homeostatic loops. The following example introduces the idea of interlocking loops.

Example 5: Blood pressure regulation. Consider the mammalian circulatory system (Fig. 2.3). Let x_1 denote the volume of blood in the arterial compartment and x_2 the volume of blood in the venous compartment. The following dynamics is simplified but will suffice for the purpose of this example:

$$\dot{x}_1 = \psi_{\text{CO}}(x_1, x_2) - \psi_{\text{RO}}(x_1, z_0) - \psi_{\text{Q}}(x_1, x_2, z) \tag{2.15}$$
$$\dot{x}_2 = u - \psi_{\text{CO}}(x_1, x_2) + \psi_{\text{Q}}(x_1, x_2, z) \tag{2.16}$$

where $u > 0$ is the net fluid input to the system (alimentary water intake plus any infusions, less non-urinary excreta), ψ_{CO} represents cardiac output, ψ_{RO} represents renal output (fluid loss through urinary excretion), and ψ_{Q} represents the blood flow through the tissues of the body. The ψ_{X}-type quantities here represent blood flows, that is, ψ_{X} is the perfusion rate of category $\text{X} \in \{\text{CO}, \text{RO}, \text{Q}\}$. From a physiological point of view, this blood flow is a function of the pressures both upstream and downstream from X. Treating these flows as functions of x_1 and x_2 is warranted since arterial pressure P_{A} is a monotone increasing function of x_1 and venous pressure P_{V} is a monotone increasing function of x_2 (spatial as well as

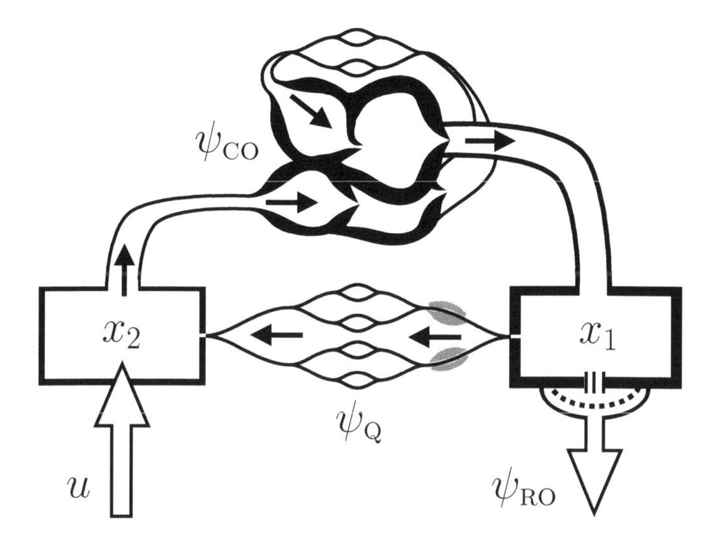

Figure 2.3 **Simplified diagram of mammalian circulatory system.** Further explanation in the text.

temporal variations exist within both the arterial and venous systems; the pressures corresponding to x_1 and x_2 are averages over the respective systems, time-averaged over a period of e.g. a minute).

The actuators z_0 and $z = [z_1, \ldots, z_n]^\mathsf{T}$ represent the smooth muscle tone in the walls of the smallest arterial vessels directly upstream from the vascular bed (the *arterioles*). The elements of z correspond to the various organs and tissues, whereas z_0 controls the input resistance to the ultrafilter function of the kidneys (this governs urine production; in terms of a metabolically active tissue with its own proper blood supply, the kidneys are separately represented as elements of z just like any other vascular bed).

Even without fully specifying the (non-linear) functions ψ_X, we can deduce a number of results. At steady state, we have $\dot{x}_1 = \dot{x}_2 = 0$ which implies $u^* = \psi_{RO}(x_1, z_0)$. If we let $\psi_{RO}^{-1}(\cdot; z_0)$ represent the inverse of ψ_{RO} with respect to its first argument, we have $x_1 = \psi_{RO}^{-1}(u^*; z_0)$ and it follows that steady-state arterial blood pressure, with z_0 fixed, is determined by net fluid intake u^*, via $P_A(\psi_{RO}^{-1}(u^*; z_0))$. This contradicts the received medical wisdom that (pathological) changes in arteriolar tone (z) play a causative role in hypertension. Similarly, occlusions or vascular remodelling in the peripheral bed, which would alter the function ψ_Q, would not affect steady-state arterial blood pressure. As long as u^* remains the same, P_A will return to the same value after perturbations, regardless of any structural alterations in ψ_{CO} or ψ_Q; unfortunately, this effect has been labelled the 'infinite gain' of the kidney-fluid mechanism for pressure control [32], which is inappropriate since the steady-state value of P_A *does* in fact vary with u^*.‡‡

On the other hand, u^* and ψ_{RO} are almost negligible compared to overall turnover of the system ψ_Q ($\psi_{RO}/\psi_Q \approx 10^{-4}$), giving $\psi_{CO}(x_1, x_2) \approx \psi_Q(x_1, x_2, z)$ as our second steady-state condition. With x_1 (and hence arterial pressure) governed primarily by u^*, this second condition determines x_2 (and hence venous pressure). If we let $\psi_{CO}^{-1}(\cdot; x_1)$ represent the inverse of ψ_{CO} with respect to its second argument, we have $x_2 = \psi_{CO}^{-1}(\psi_Q(x_1, x_2, z); x_1)$ at steady state. Given that ψ_Q depends only weakly on x_2, we can take this last equation as stating how the steady-state value of x_2 varies with x_1. Moreover, since ψ_{CO} does not vary strongly with x_1, the dependence of x_2 on x_1 at steady-state runs primarily through the function ψ_Q. It follows that structural changes in the vascular bed, which alter ψ_Q as a function of x_1, z, and (weakly) x_2, will affect venous blood pressure, as opposed to arterial blood pressure, which, as we have seen, remains unaltered as long as fluid intake remains nominal.

All conclusions up to this point were obtained on the assumption that the actuator variables remain fixed. Since steady-state arterial blood pressure depends on u^* and z_0, and the former is treated as a forcing function, we see that adjustment of z_0 can regulate P_A and render it independent of u^* at steady state. To illustrate this, let us consider the following simple dynamics for z_0:

$$\dot{z}_0 = \eta_0 \left(\widehat{P}_A - P_A(x_1(t)) \right) \tag{2.17}$$

where $\eta_0 > 0$ is a rate-governing parameter and $\widehat{P}_A$ the apparent set-point of the system (which we feel justified in introducing as such, in view of the discussion in Section 2.1; that is, we understand that this set-point is a compound conglomerate of properties of the biological components that make up the control loop, which in the present case involve the central nervous system [63]). This expression assumes $\partial\psi_{RO}/\partial z_0 < 0$ as an increase in arteriolar smooth muscle tone leads to an increase of flow resistance. Equation (2.17) is another example of integrating control, where the 'error' $\widehat{P}_A - P_A$ is integrated to determine the control output z_0

‡‡One may nonetheless be impressed that the steady-state value of P_A does not vary with the perturbations noted, and call the reciprocal of this zero slope the 'gain.' However, this overlooks the fact that the control loop at hand does not have the character of proportional control, but rather of integrating control, as x_1 effectively acts as its own error integrator.

(recall that 'error' is to be interpreted without connotations of intention and purpose).

The elements of z, corresponding to the various vascular beds, do not individually exert a strong influence on P_A. Each does, however, strongly determine the blood supply to the corresponding bed. Therefore it is reasonable to suppose that each z_i has dynamics driven by the adequacy of the blood supply. The model would thus have to be extended to represent the stock of e.g. oxygen or ATP in the tissue served. Let us refrain from giving detailed equations and just assume the (plausible) outcome that arteriole tension for each vascular bed is adequately and rapidly adjusted to the requirements of the tissue served.

Referring back to the venous steady-state condition, we now appreciate that, taking adjustment of z into account, ψ_Q will be regulated to requirements at the organismal level. It follows that P_V is *effectively uncoupled* from P_A at steady-state: the latter would vary only with fluid intake, but adjustments in the input resistance of the ultrafilter (z_0) achieve regulation to $\widehat{P}_A$, whereas venous blood pressure is not regulated as such but enslaved to total perfusion ψ_Q. In fact, this constitutes a *second pressure-fluid effect*, whereby the strong dependence of cardiac output ψ_{CO} on x_2 allows automatic alignment of cardiac output and organism-level perfusion ψ_Q.

This analysis ignores a slew of additional phenomena, such as acute central regulation of cardiac output by varying heart rate and strength of contraction, additional mechanisms safeguarding minimal pressure in the largest venous vessels, the role of the lymphatic system, the relationship between arterial elastance, pulse wave velocity, impedance matching and cardiac performance, and so on. While important, these effects all act as modulators of the primary control loops which we have sketched here. The take-home message for the purposes of the present monograph is that the essential regulatory structure is based on interlocking control loops regulating arterial blood pressure and adequate tissue perfusion. The loops are affected by shared actuator variables.

Our qualitative analysis has proceeded quite far in the absence of precise specifications of the functions ψ_X. Our results only depend on general ideas regarding the sense and strength of the partial derivatives of the ψ_X-functions. As long as these qualitative characteristics remain the same, changes in the properties ('shapes') of these functions, even as a result of local pathologies, should *not* change the ability of the system to regulate. (The case $u^* < 0$ — bleeding out — would constitute a qualitative change where we should expect the above homeostatic loops to fail; but if a bleed, even a serious one, is reversed and $u^* > 0$ restored, we still expect to return to normal.) This insight can hardly be novel, although it proves difficult to find in the literature and the common wisdom tends to veer in the opposite direction, where changes in vascular properties (resting diameter and stiffness, which affect rheological characteristics such as complicance, resistance, and impedance) are viewed as primary causes rather than secondary responses to the underlying disease.

In any event, it would appear that blood pressure regulation is extremely robust, even in the face of sundry vascular pathologies. The first pressure-fluid effect protects mean arterial blood pressure as long as there are no renal pathologies affecting ultrafiltration, and the second pressure-fluid effect adjusts cardiac output to demand, as long as there are no advanced cardiac pathologies.

However, we have tacitly been supposing that the actuator variables can assume any value that might be required by continued homeostasis. The biological reality is that the arteriolar myocytes can range between fully relaxed and fully constricted, and this defines a finite, bounded range of vascular flow resistance values. In other words, each actuator is subject to a constraint of the form

$$z_{j,\min} \leq z_j \leq z_{j,\max} \tag{2.18}$$

and when one or more of these 'hits the stops,' failure of compensation may occur. This may well be a general principle of ætiology: pathological processes change the properties of the physiological component, but the regulatory (actuator) component will, at least in the course of normal operation, compensate and keep the alterations from manifesting as system dysfunction, until a point is reached where such compensation would require one or more actuator variables to take on physiologically inaccessible values, and the illness suddenly manifests itself in a possibly catastrophic manner. The underlying changes ought to be observable before this catastrophic shift happens; that is, diseases of the type we have in mind here should be diagnosable whilst still subclinical. ❖

In this example, as well as in earlier ones, we described the behaviour of actuator ($\mathcal{Z}\sim$) variables on the basis of physiological intuition. This is somewhat unsatisfactory on two counts: first, intuition may lack intersubjectivity, and, second, systems of arbitrarily high complexity may not be amenable to a freewheeling intuitive approach. A more objective and systematic procedure is wanted.

Moreover, the example suggests that interesting phenomena, such as qualitative changes in behaviour, occur when $\mathcal{Z}\sim$variables approach boundaries of their respective sets of allowed values. Although there is no *a priori* restriction on the structure of the set of values that is available to a $\mathcal{Z}\sim$variable, the typical case is that of an interval of the form given in Eq. (2.18), or perhaps a union of such intervals; in most cases fairly straightforward physico-chemical considerations suffice to warrant this assumption (e.g., a permeability cannot be negative, the outer surface of a cell cannot become too crowded with ion channels, and so on). Furthermore, there may be a value in the allowed interval that is favoured *ceteris paribus* (or several such values, or subintervals).

In sum, our considerations up to this point suggest that we need a systematic approach to indicate how $\mathcal{Z}$~variables respond to perturbations in the $\mathcal{X}$~variables, and a way of accommodating the pressures exerted on $\mathcal{Z}$~variables by the 'stops' at the end points of allowed intervals of values. These desiderata are met by the gradient-driven dynamics approach described in the following chapter.

is the *Pasteur effect* [9]. Figure. 3.2 (middle panel) shows a variation on the XOR-theme in which z_1 and z_2 are mutually compensatory, the minimum of L lying along a line in the (z_1, z_2)-plane that expresses a linear constraint. In Fig. 3.2 (right-most panel), finally, z_1 and z_2 are not allowed to both be high at the same time, implementing the Sheffer stroke or 'NAND' [46].

3.2 THE QUASI-STATIONARY CO-STATE APPROXIMATION

To obtain the reduced dynamics $\dot{\boldsymbol{x}} = \widetilde{f}(\boldsymbol{x}, \boldsymbol{u})$, Eq. (2.9), we seek a function $\overline{\boldsymbol{g}}$ that gives $\boldsymbol{z}$ as a function of $\boldsymbol{x}$ and $\boldsymbol{u}$. This function $\overline{\boldsymbol{g}}$ can be constructed on the basis of the physiological dynamics $\boldsymbol{f}$ and the physiological Lagrangian L, by exploiting optimal control principles. In particular, Pontryagin's principle identifies putative optimal control trajectories associated with the dynamics $\boldsymbol{f}$ and the objective of minimising J as defined by Eq. (3.1) [69]. Such trajectories consist of segments (*arcs*) which can be either *bang-bang* or *singular* [69]. In bang-bang control, one or more control variables is set to a boundary point of an interval of allowed values for the control variable concerned (i.e., $\boldsymbol{u} \in \text{extr}(\mathcal{U})$). Adding Anschlag terms to L for such control variables, we can force its value to be internal to that interval, but within a distance ϵ of the boundary. In this manner, all bang-bang segments can be approximated by singular segments.[‡] We assume that the trajectories obtained under this perturbation will converge (at least point-wise) to the unperturbed system's bang-bang trajectories under the limit $\epsilon \to 0$. The upshot is that we can disregard the bang-bang possibility altogether, and restrict our attention to singular control.

We can treat $\boldsymbol{z}$ as the singular control variable if there are no range limitations on the elements of $\dot{\boldsymbol{z}}$ and the elements of $\boldsymbol{z}$ are all 'ϵ-limited' via Anschlag terms. The absence of range limitations on the rates of change of the elements of $\boldsymbol{z}$ presupposes that the $\mathcal{Z}\sim$variables can change

[‡]Perturbation *away* from singular arcs is also possible; this has the effect of turning the entire control into bang-bang control; that is, L can be modified slightly so as to turn all arcs, including the singular ones, into arcs with extremal controls [8]. This move is motivated by (i) the *Bang-Bang principle*, which states that the set of available control values $\mathcal{U}$ can be reduced to the set of its extremal points extr($\mathcal{U}$) without affecting the set of state values that can be attained at any given finite time [6], and (ii) the substantial reduction in computational (algorithmic) difficulty if one may assume that singular arcs are absent [8].

arbitrarily fast. If this assumption is not warranted, a different approach is required, which we pursue in Section 3.4.

For the sake of simplicity, we assume that the external input $\boldsymbol{u}$ is time-constant, i.e. $\boldsymbol{u}(t) \equiv \overline{\boldsymbol{u}}$. The Pontryagin-Hamilton function H associated with the control problem is a Legendre transform of the physiological Lagrangian [69]:

$$H(\boldsymbol{\lambda}, \boldsymbol{x}, \boldsymbol{z}, \overline{\boldsymbol{u}}) = L(\boldsymbol{x}, \boldsymbol{z}) + \langle \boldsymbol{\lambda}, \boldsymbol{f}(\boldsymbol{x}, \boldsymbol{z}, \overline{\boldsymbol{u}}) \rangle \tag{3.3}$$

where $\boldsymbol{\lambda}$ is the co-state with $\dim(\boldsymbol{\lambda}) = \dim(\boldsymbol{x}) = n$ and $\langle, \cdot, \cdot \rangle$ denotes the inner product. Hamilton's equations state that

$$\dot{\lambda}_i = -\frac{\partial H}{\partial x_i} = -\frac{\partial L}{\partial x_i} - \langle \boldsymbol{\lambda}, \frac{\partial \boldsymbol{f}}{\partial x_i} \rangle \tag{3.4}$$

for $i = 1, \dots, n$. Furthermore, singular control means that the *switching function* vanishes for all z_j, $i = 1, \dots, m$:

$$\frac{\partial H}{\partial z_j} = \frac{\partial L}{\partial z_j} + \langle \boldsymbol{\lambda}, \frac{\partial \boldsymbol{f}}{\partial z_i} \rangle = 0\,. \tag{3.5}$$

Equations (3.4) and (3.5) can be represented in matrix form:

$$(\nabla_{\boldsymbol{x}} \boldsymbol{f}) \cdot \boldsymbol{\lambda} = -\nabla_{\boldsymbol{x}} L - \dot{\boldsymbol{\lambda}} \quad \text{and} \quad (\nabla_{\boldsymbol{z}} \boldsymbol{f}) \cdot \boldsymbol{\lambda} = -\nabla_{\boldsymbol{z}} L \tag{3.6}$$

where $\nabla_{\boldsymbol{x}} \boldsymbol{f}$ is an $n \times n$ matrix whose transpose $(\nabla_{\boldsymbol{x}} \boldsymbol{f})^{\top}$ is the Jacobian matrix of the dynamics $\boldsymbol{f}$ (with $\boldsymbol{u}$ and $\boldsymbol{z}$ considered as fixed, independent variables for the purposes of differentiation) and $\nabla_{\boldsymbol{z}} \boldsymbol{f}$ is an $m \times n$ matrix. When $\nabla_{\boldsymbol{x}} \boldsymbol{f}$ is invertible, we may combine these equations as follows:

$$\nabla_{\boldsymbol{z}} L = (\nabla_{\boldsymbol{z}} \boldsymbol{f}) \cdot (\nabla_{\boldsymbol{x}} \boldsymbol{f})^{-1} \cdot \left(\nabla_{\boldsymbol{x}} L + \dot{\boldsymbol{\lambda}} \right). \tag{3.7}$$

The *co-state equilibrium condition* is obtained by setting $\dot{\boldsymbol{\lambda}} = \boldsymbol{0}$:

$$\nabla_{\boldsymbol{z}} L = (\nabla_{\boldsymbol{z}} \boldsymbol{f}) \cdot (\nabla_{\boldsymbol{x}} \boldsymbol{f})^{-1} \cdot \nabla_{\boldsymbol{x}} L \tag{3.8}$$

which furnishes a necessary condition for the function $\overline{\boldsymbol{g}}$, provided that $\dot{\boldsymbol{\lambda}}$ stays 'sufficiently close' to $\boldsymbol{0}$. The latter requirement is essential; we shall refer to it as the *quasi-stationary co-state (QSC) approximation.*

Under the prevailing $\overline{\boldsymbol{u}}$, the system may settle on a steady state in which all $\mathcal{X}$~variables reside at their physiological optimum, this being a state of global satisfaction. Such a state is characterised by $\nabla_{\boldsymbol{x}} L = \boldsymbol{0}$, which implies that $\nabla_{\boldsymbol{z}} L = \boldsymbol{0}$.

However, there may not be an allowed value of $\boldsymbol{z}$ that permits global satisfaction. In that case, several of the $\mathcal{X}$~variables are displaced from their respective optima, in what we may call a *frustrated state*. If we allow $\boldsymbol{u}(t)$ to change, but do so sufficiently slowly, i.e. to allow the system state $(\boldsymbol{x}, z)$ to remain close to the value pertaining to $\boldsymbol{u}(t)$ if the latter were held constant indefinitely, Eq. (3.8) remains a valid condition on z. This defines a process similar to the quasi-static processes encountered in thermodynamics, and also somewhat analogous to adiabatic processes in quantum mechanics [31].

Example 7: temperature control (continued). Let us revisit Example 2, where $n = m = 1$. We posit the following:

$$L = \frac{(x-1)^2}{2} + \eta\frac{(z-z_0)^2}{2} \tag{3.9}$$

where z_0 and $\eta > 0$ are fixed parameters. To apply Eq. (3.8), we first compute $\nabla_z L = \eta(z - z_0)$, $\nabla_{\boldsymbol{x}} L = x - 1$, and $\nabla_z f = +1$. Assembling the pieces, we find

$$z = z_0 + (1-x)/\eta \tag{3.10}$$

in agreement with the steep section of $\overline{g}(x)$ in Fig. 2.2. For constant input u^*, x will tend to the value $(1-\eta(u^* - z_0))/(1-\eta)$, which is arbitrarily close to 1 as we take η down to zero.

As yet, the flat sections of the graph of $\overline{g}(x)$ in Fig. 2.2 are missing, because L, as specified by Eq. (3.9), lacks the Anschlag terms. If we add such terms ('Anschläge'), as described by Eq. (3.2), we find that Eq. (3.10) remains valid, provided that $\min\{z - z_{\min}, z_{\max} - z\} > \epsilon$. Otherwise, the balance between the homeostatic pressure L_x and the stop pressure L_z is dominated by the steep (diverging) gradient of the Anschlag term that prevails within a distance ϵ of the end points. Moreover, in the double limit $\eta \to 0$, $\epsilon \to 0$, the behaviour of the model converges to that of classical *bang-bang* control: here $z = z_{\min}$ when $x > 1$, $z = z_{\max}$ when $x < 1$, and, finally, $z = z^* = 1 - u^*$ when $x = 1$. In this context z^* is called the *singular control value*.[††] ❖

Example 8: temperature control with explicit hormones. Let us take the same model as in the previous example but suppose that we know that the actuator variable is proportional to the concentration of a

[††]The conventional optimal-control approach to the problem does not deploy Anschläge and instead imposes the constraints on z directly. One then derives bang-bang arcs followed by the terminal singular arc. The selection of either $z_{\min}$ or $z_{\max}$ during bang-bang control is determined by the sign of the switching function [69].

certain hormone, which follows first-order kinetics:

$$\dot{x}_1 = u - x_1 + x_2 \tag{3.11}$$

$$\dot{x}_2 = z - \lambda x_2 \tag{3.12}$$

where λ is a positive constant. The new actuator variable z is the secretion rate of the hormone, and the old actuator variable has become state variable x_2. Thus, although we would be inclined to ascribe a quantity such as a hormone concentration to the regulatory component $\boldsymbol{x}_{\text{reg}}$, there is no harm in transferring any known parts of the latter to the physiological component $\boldsymbol{x}$ (the latter is just that portion of the integrated physiology of the organism for which we are able and willing to write down explicit dynamics). We let

$$L = (x_1 - 1)^2/2 + \sigma(z)$$

where $\sigma(\cdot)$ contains the Anschlag terms as in Eq. (3.2). Specifying the terms in Eq. (3.8), we find

$$\sigma'(z) = \begin{bmatrix} 0 & 1 \end{bmatrix} \cdot \begin{bmatrix} -1 & 0 \\ -\lambda^{-1} & -\lambda^{-1} \end{bmatrix} \cdot \begin{bmatrix} x_1 - 1 \\ 0 \end{bmatrix} = \lambda^{-1}(1 - x_1) \tag{3.13}$$

which can be solved for a unique z-value, since $\sigma'(\cdot)$ is monotone. Secretion of the hormone, z, 'bangs' between (ϵ-neighbourhoods of) the minimum rate and the maximum rate, according to whether $x_1 > 1$ or $x_1 < 1$. The minimum secretion rate of a hormone can be assumed to equal zero, corresponding to lack of secretory activity.

The stability of the equilibrium point can be analysed by inspecting the eigenvalues of the Jacobian matrix of $\widetilde{\boldsymbol{f}}$ (not to be confused with $(\nabla_{\boldsymbol{x}} \boldsymbol{f})^{\mathsf{T}}$ which is the Jacobian matrix of $\boldsymbol{f}$, evaluated with $\boldsymbol{z}$ held constant). These eigenvalues are not defined for the present system; instead, we consider the behaviour of the system with the following physiological Lagrangian:

$$L = L = (x_1 - 1)^2/2 + \sigma(z) + \eta\,(z - z_0)^2/2 \tag{3.14}$$

where z_0 is positive but smaller than the maximum secretion rate, and study the limit $\eta \to 0$. We find that the real parts of both eigenvalues are strictly negative and that the approach to the equilibrium by the local linearised system is a stable spiral with an oscillation frequency that diverges as $\eta \to 0$. This phenomenon is known as *chattering*.

If we retain a positive η in Eq. (3.14), we find that the chattering phenomenon is avoided. Moreover, if η is set sufficiently high, the oscillatory approach to equilibrium is abolished altogether. However, this comes at a price: the equilibrium is no longer located exactly at $x_1 = 1$, but displaced, since the term $\eta(z - z_0)^2/2$ introduces a homeostatic optimum for the actuator variable z at z_0, and the final state represents a 'frustrated' compromise, as we saw in Example 7. ❖

A pure feedback form was found in both Examples 7 and 8, i.e. $\boldsymbol{z}$ was found to be a function of $\boldsymbol{x}$ alone. However, the QSC formalism also allows pure feedforward, and more generally for mixed feedback/feedforward. While we have not introduced L as explicitly dependent on the input $\boldsymbol{u}$, the dynamics $\boldsymbol{f}$ does depend on $\boldsymbol{u}$, and in this way the input may enter the QSC condition that relates the gradients $\nabla_{\boldsymbol{x}} L$ and $\nabla_{\boldsymbol{z}} L$.

Example 9: hæmolymph sodium concentration in an aquatic invertebrate (continued). In Eq. (2.11) from Example 3, let us take $\beta = \gamma = 0$ and treat α as an actuator variable, which we shall write as $z > 0$. Thus:

$$\dot{x} = z(u - x) \tag{3.15}$$

where we recall that x represents the hæmolymph sodium concentration, u the sodium concentration in the animal's environment, and $z > 0$ the permeability of the animal's integument. From Eq. (3.8) we deduce $zL_z = (x - u)L_x$, where subscripts denote partial derivatives. Whether permeability z is set to a low or a high value now depends not only on whether x is above or below 1, but also on whether u is above or below x. This is an example of a mixture of feedback and feedforward control.

From a physiological point of view, we observe that an increase in permeability expedites a return to $x = 1$ when $u > x$ and $x < 1$; but when $u < x$ and $x < 1$, only a further decrease of x is possible and the best that can be done is to minimise the rate at which sodium is lost, by reducing skin permeability to $[Na^+]$. To extend the example, we recall that the organism's skin contains active sodium transporters, which can pump sodium ions into the hæmolymph and thus help the animal maintain $x = 1$ in an environment where $u < 1$ [94]. Thus:

$$\dot{x} = z_1(u - x) + z_2 \tag{3.16}$$

where z_1 is integumental permeability, corresponding to z in Eq. (3.15), and $z_2 \geq 0$ denotes the rate of active sodium uptake.

From Eq. (3.8) we have $z_1 L_{z_1} = (x - u)L_x$, as before, along with $L_{z_2} = L_{z_1}/(u - x)$, which shows that for $x > u$ the actuators must be at opposite Anschläge, e.g. high sodium influx z_2 along with minimal integumental sodium permeability z_1. ❖

3.3 THE MARČENKO-PASTUR EFFECT

The actuators correspond to physiological parameters (fluxes, permeabilities, expression levels and degrees of activation of functional proteins such as enzymes, ion channels, etc.) and we should therefore allow the non-negative entries of the matrix $\nabla_z \boldsymbol{f}$ to occur in whichever way they will. That is to say, we are not in a position to impose any *a priori* constraints, e.g., demanding that $\nabla_z \boldsymbol{f}$ be sparse or diagonal (or, failing that, diagonally dominant). This freedom is why we say that the control loops are generically *interlocking*.[‡‡]

However, such freedom does raise the concern that the QSC approximation gives rise to destabilised system dynamics. It turns out that such concerns are mostly unwarranted and that the dynamics under QSC will not be any worse than is already occasioned by the stability structure of $\nabla_{\boldsymbol{x}} \boldsymbol{f}$ (which signifies the intrinsic local dynamics of the physiological component). This is the *Marčenko-Pastur effect*, which strictly speaking, only obtains for arbitrarily large systems; however, a little numerical experimentation will convince the reader that even for systems of modest size the Marčenko-Pastur effect is already in force.

Example 10: randomly connected control. Let us generalise Example 7 to the general case $n = m > 1$. With $\boldsymbol{x} \in \mathbb{R}^n$, $\boldsymbol{u} \in \mathbb{R}^n$, and $\boldsymbol{z} \in \mathbb{R}^n$, we consider the following system:

$$\dot{\boldsymbol{x}} = \boldsymbol{u} - \boldsymbol{x} + \mathbf{K} \cdot \boldsymbol{z} \tag{3.17}$$

$$L = \sum_{i=1}^{n} \frac{(x_i - 1)^2}{2} + \eta \sum_{i=1}^{n} \frac{(z_i - z_{i,0})^2}{2} \tag{3.18}$$

where $\mathbf{K}$ denotes an $n \times n$ matrix containing the *coupling coefficients* (here, $\nabla_z \boldsymbol{f} = \mathbf{K}^\top$). If we let $\xi_i = x_i - 1$ and $\zeta_i = z_i - z_{i,0}$, the QSC approximation gives:

$$\dot{\boldsymbol{\xi}} = \boldsymbol{u} - \left(\mathbf{I} + \eta^{-1} \mathbf{K} \cdot \mathbf{K}^\top\right) \cdot \boldsymbol{\xi}\,. \tag{3.19}$$

The matrix $-\mathbf{I} - \eta^{-1} \mathbf{K} \cdot \mathbf{K}^\top$ governs the stability of this system; an obvious and robust way of achieving this would be to demand that $-\mathbf{K} \cdot \mathbf{K}^\top$ be stable. By the Spectral Theorem, the eigenvalues of $\mathbf{K} \cdot \mathbf{K}^\top$ are certainly real, so for stability we must require that they are all positive. This

[‡‡] A physiology that leaves $\nabla_z \boldsymbol{f}$ diagonal is not impossible in principle. We could 'diagonalise' the physiological component itself, by repeatedly applying the technique of splitting an actuator into two separate actuators, interposing a physiological entity such as an additional compartment, buffer, reserve, reaction step, etc.

is almost surely the case as n becomes arbitrarily large, for the eigenvalue spectrum then converges (with Kolmogorov distance $O(1/\sqrt{n})$ in probability [28]) to the Marčenko-Pastur distribution law [52]. Although the eigenvalues of $\mathbf{K} \cdot \mathbf{K}^{\mathrm{T}}$ are all non-negative by the Marčenko-Pastur theorem, some of them will be quite small.

Thus, if we form links between the actuators and the physiological state variables in a probabilistic manner (the theorem requires independent, identically distributed random variables, but it may be surmised that independence is the critical criterion here), we should expect the system to be able to regulate and attain near-global satisfaction (the parameter η negotiates the trade-off between $\mathcal{X}\sim$ and $\mathcal{Z}\sim$variables), particularly if the system is large — this is, incidentally, one of several instances in theoretical biology where more 'complex' systems can actually be 'simpler' in one crucial respect.

An immediate corollary of biological interest is that evolution disposes of a substantial amount of leeway when we think of it as 'random tinkering' guided and corrected by natural selection. Another corollary is that we are almost assured that the regulated dynamics inherits stability problems from the physiological component; that is, if we generalise Eq. (3.17) to the following:

$$\dot{x} = u - \mathbf{S} \cdot x + \mathbf{K} \cdot z \tag{3.20}$$

where $\mathbf{S}$ is a *system matrix*, then the similarly generalised version of Eq. (3.19) will inherit the stability properties of $\mathbf{S}$.

Although this example is linear, the conclusions regarding the Marčenko-Pastur effect carry over to the local, linearised, analysis of general non-linear systems. ❖

This example throws new light on our main theme of interlocking control loops. Interlocking is completely absent only when both $\mathbf{M}$ and $\mathbf{K}$ in Eq. (3.17) are diagonal. The Marčenko-Pastur effect concerns $\mathbf{K}$ and tells us that the 'randomly connected' system will generally perform just as well as a system whose $\mathbf{K}$ is diagonal. This prompts us to ask what, if anything, is special about interlocking control loops at the level of $\mathbf{K}$.

If a non-zero element in a diagonal $\mathbf{K}$ is set to zero (corresponding to the severing of a functional link in the control loop), regulation of the corresponding $\mathcal{X}\sim$variable is completely lost. By contrast, setting a non-zero element of a 'random' $\mathbf{K}$ to zero has a vanishingly small chance (in the limit $n \to \infty$) of disrupting the regulative properties of the system. This suggests that interlocking control loops are more robust, both in the evolutionary sense (gradual adaptation is possible) and the life-history sense (a lesion in one pathway is compensated by *redundancy*, as expressed by the number of non-zero elements in **K**).

Example 10 lacks 'stops' (i.e., limitations on the values that the actuators may attain) and the Marčenko-Pastur effect tells us that intricate (indeed, seemingly random) coupling does not diminish regulatory capability. This reinforces the idea that such stops are essential in understanding loss of regulation or compensation.

3.4 DYNAMIC REGULARISATION OF THE ACTUATOR VARIABLES

If we wish to treat z as a classic control variable, we must in general forgo the option of imposing constraints on the rate of change $\dot{z}$. In reality, there will always be natural limits on how rapidly the actuator variables are capable of being adjusted. Nevertheless, as long as the physiological variables change much more slowly, i.e. $\max_{i=1,\ldots,n}\{\dot{x}_i\} \ll \min_{j=1,\ldots,n}\left\{\dot{z}_j\right\}$, we may treat $|\dot{z}|$ as essentially unbounded. When this time-scale argument fails, we could promote the actuator to the physiological component; i.e., we turn any 'slow' $\mathcal{Z}$~variables into $\mathcal{X}$~variables, which necessitates the introduction of one or more new $\mathcal{Z}$~variables to describe the dynamics of this newly created $\mathcal{X}$~variable, as shown in Example 8.

Two values of z (or even more) may satisfy Eq. (3.8) for a given pair $(\boldsymbol{x}, z)$. This could be resolved by introducing a dynamic 'flag' variable that indicates the solution branch along which z is moving. Moreover, the reduced dynamics $\widetilde{\boldsymbol{f}}$ may inherit adverse dynamic properties from $\boldsymbol{f}$, as we saw in Example 10. These technical difficulties can be circumvented if we think of Eq. (3.8) as the equilibrium condition for a dynamical law on $\mathcal{Z}$, such as the following:

$$\dot{z} \quad = \quad -\begin{bmatrix} \mu_1 & & 0 \\ & \ddots & \\ 0 & & \mu_m \end{bmatrix} \cdot \left(\nabla_z L - (\nabla_z \boldsymbol{f}) \cdot (\nabla_x \boldsymbol{f})^{-1} \cdot \nabla_x L\right) \tag{3.21}$$

where $\mu_1, \ldots, \mu_m$ are positive rate constants. Here, $\boldsymbol{x}$ and z are coupled in a manner that is liable to evoke oscillations. If we suppose that L depends on the z_j via the Anschlag function, Eq. (3.2), and we consider the limit $\mu_j \to \infty$ for all $j \in \{1, \ldots, m\}$, we see that each of the z_j relaxes with an arbitrarily small time constant $1/\mu_j$ toward either its lower or upper stop, depending on the signs of the elements of the gradient $\nabla_{\boldsymbol{x}} L$. Only when $\nabla_{\boldsymbol{x}} L = \mathbf{0}$ are all of the z_j free to take on values in the corresponding intervals $[z_{j,\min} + \epsilon, z_{j,\max} - \epsilon]$. As $\boldsymbol{x}$ approaches a state such that $\nabla_{\boldsymbol{x}} L = \mathbf{0}$, it will oscillate about this state with diverging frequency and diminishing amplitude, while the $\mathcal{Z}$~variables exhibit chattering toward the singular value prescribed by Eq. (3.8).

Example 11: temperature control (continued). In Example 7, we worked with

$$L = \frac{(x-1)^2}{2} + \eta \frac{(z - z_0)^2}{2} + \text{Anschlag} \tag{3.22}$$

where $z_0, \eta > 0$. We have indicated that an Anschlag as in Eq. (3.2) is to be added. If z is governed by Eq. (3.21), we obtain

$$\begin{aligned} \dot{x} &= u - x + z \\ \dot{z} &= -\mu \left(\frac{\partial L}{\partial z} + x - 1 \right) \end{aligned} \tag{3.23}$$

where $\mu > 0$ is a rate parameter characterising the regulatory subsystem.

For $\partial L / \partial z = 0$, this system describes integrating control [39], as we obtain $\dot{z} \propto x - 1$, which represents the 'error' or deviation from the optimum value 1. For the standard Anschlag, Eq. (3.2), we have a nil contribution to $\partial L / \partial z$ if $z \in [z_{\min} + \epsilon, z_{\max} - \epsilon]$. Setting $\eta = 0$ and applying a constant input $u \equiv u^*$, we deduce that the system tends toward the state $(x, z) = (1, z^*)$ where $z^* = 1 - u^*$ is the singular value, provided that $z^* \in [z_{\min} + \epsilon, z_{\max} - \epsilon]$; otherwise the steady state will feature z stopped at a lower or upper bound, with $x \neq 1$. For $0 < \mu \leq 1/4$, approach to this equilibrium point is non-oscillatory, and for $\mu \to 0$, one of the modes will become arbitrarily slow. For $\mu > 1/4$, the system exhibits damped oscillation, with a frequency that diverges as $\mu \to \infty$.

In the present example, to obtain integrating control in the classical sense of the word, we must set $\eta = 0$, but for η merely 'small enough' we expect behaviour quite similar to integrating control; the parameter η negotiates the balance between the homeostatic drive toward $x = 1$ and the drive that favours $z = z_0$. ❖

This example indicates that Eq. (3.21) is a generalised version of integrating control, in keeping with the claim, which has been put forward by several authors, that integrating control is ubiquitous in biological regulation [19, 29, 99]. We obtain the (rather common) special case of classic integrating control when the leading term in the local expansion of L about the optimum point happens to be of the lowest possible order (i.e., 2).

The reduced model $\widetilde{f}$, with z as per Eq. (3.8), or as a solution of the ODE, Eq. (3.21), will generically inherit any stability issues associated with the original dynamics f. This can be addressed by subtracting a diagonal matrix:

$$\mathbf{M} = \nabla_x f - \begin{bmatrix} \varrho_1 & & \\ & \ddots & \\ & & \varrho_n \end{bmatrix} \tag{3.24}$$

where $\varrho_1, \ldots, \varrho_n$ are positive constants, chosen so as to render $\mathbf{M}$ stable and its diagonal elements all negative, which is always possible in virtue of the Bauer-Fike theorem [7]. The matrix $\mathbf{M}$ replaces $\nabla_x f$ in Eq. (3.21).

As in Example 3, we should take care not to mistake our notational devices for the inner workings of the biological regulatory system. The danger is that our calculations are taken to suggest that the biological system draws up matrices, transposes or inverts them, and so on. As a mechanism, the system at hand integrates afferent (input) signals and converts them into efferent (output) signals that set the values of the actuators. The inputs are represented, in the present system of notation, as gradients ($\nabla_x L$ for physiological 'requirements' and $\nabla_z L$ for actuator constraints), whereas we represent the information processing by linear operators, not because they are the only possible modelling choices, but because they appear suitable for the purpose of developing the theory in the simplest possible way.

However, this does raise the question of why we should expect that natural selection ends up at forms equivalent to the ones we have been deriving. The fortuitous fact is that *some* degree of effective control is attained provided that $\dot{z}_j$ correlates negatively with $(\partial f_i/\partial z_j)(\partial L/\partial x_i)$ for all i [14]. Evolution thus acquires a viable substratum for fine-tuning these correlations and our calculations link minimisation of $J = \int L dt$ to the couplings as stated in Eq. (3.8) — which is exact under the QSC approximation.

Summarising the development in the present chapter, we have proposed a dynamics on the state space $\mathcal{X} \times \mathcal{Z}$, with the following general form:

$$\begin{cases} \dot{x} & = f(x, z, u) \\ \dot{z} & = -\mathbf{D}_\mu \cdot \nabla_z L + (\nabla_z f) \cdot \mathbf{M}^{-1} \cdot \nabla_x L \end{cases} \tag{3.25}$$

where $\mathbf{D}_\mu$ denotes the diagonal matrix containing the $\mu_1,\ldots,\mu_m$ in Eq. (3.21) and $\mathbf{M}$ was defined in Eq. (3.24).

The function f is assumed to be specifiable from physiological knowledge in the widest sense of the term. The function L, which we have called the physiological Lagrangian, encodes information about homeostatic drives, which we ultimately hope to connect to selective pressures, a thorny problem we will defer to Chapter 7. The matrix $(\nabla_z f) \cdot \mathbf{M}^{-1}$ encodes how these pressures, represented by $\nabla_x L$, are transmitted to the regulatory subsystem.

CHAPTER 4

Coupling and pleiotropy

We have thus far explored the strategy of dividing the state variables of a given biological system with features of (homeostatic) regulation into two classes, $\mathcal{X}\sim$variables and $\mathcal{Z}\sim$variables, such that we have physiologically 'explicit' dynamics for the former and a more abstract gradient-driven dynamics for the latter, as shown in Eq. (3.25). The division between $\mathcal{X}\sim$variables and $\mathcal{Z}\sim$variables is somewhat arbitrary, mirroring the fuzzy boundary in the division between physiology and (neuro)endocrinology. We include in the $\mathcal{X}\sim$components those aspects about which we wish to be explicit — always provided that our knowledge and data sets allow us to be explicit. Conservation and scaling laws, stoichiometric principles and the like generally go a long way toward the specification of the dynamics $\boldsymbol{f}$ of $\mathcal{X}\sim$.

In the previous chapter, we observed that the matrix $(\nabla_z \boldsymbol{f}) \cdot \mathbf{M}^{-1}$ expresses how regulatory pressures are transmitted. Put differently, $(\nabla_z \boldsymbol{f}) \cdot \mathbf{M}^{-1}$ contains the information needed to translate physiological requirements into regulatory actions. Visualising this connection in a graph-theoretical way, we can obtain a direct insight into the origin of pleiotropic suites.

4.1 THE COUPLING STRUCTURE

Let us introduce the *coupling matrix* $\boldsymbol{\gamma}$ by defining $\gamma_{ii} = 0$ and

$$\gamma_{ij} = -\frac{\partial f_i / \partial x_j}{\mathrm{M}_{ii}} \tag{4.1}$$

for $i \neq j$. By construction, we have $\text{sgn}\{\gamma_{ij}\} = \text{sgn}\{\partial f_i/\partial x_j\}$ since the constants $\varrho_1, \ldots, \varrho_n$ were chosen so as to render $\mathrm{M}_{11}, \ldots, \mathrm{M}_{nn}$ negative. Also, let $\mathbf{D}_\mathbf{M}$ be a diagonal matrix whose diagonal elements agree with those of $\mathbf{M}$, and let

$$\boldsymbol{\kappa} = -\mathbf{D}_\mu \cdot (\nabla_z f) \cdot \mathbf{D}_\mathbf{M}^{-1} \,. \tag{4.2}$$

We have $\text{sgn}\{\kappa_{ij}\} = \text{sgn}\{\partial f_j/\partial z_i\}$. The dynamics of z, Eq. (3.21), can be written in the following form:

$$\dot{z} = -\mathbf{D}_\mu \cdot \nabla_z L - \boldsymbol{\kappa} \cdot \left(\textstyle\sum_{\ell=0}^{\infty} \boldsymbol{\gamma}^\ell\right)^\mathsf{T} \cdot \nabla_x L \,, \tag{4.3}$$

where we have used the Von Neumann series for the matrix inverse, which is valid provided that $\lim_{n\to\infty} \boldsymbol{\gamma}^n = \mathbf{0}$. Calculations can be simplified by truncating this expansion. If, for instance, we take the zeroth-order truncation, we obtain

$$\dot{z} = -\mathbf{D}_\mu \cdot \nabla_z L - \boldsymbol{\kappa} \cdot \nabla_x L \,, \tag{4.4}$$

which is the Ansatz used in ref. 14.

To interpret this sum $\sum_{\ell=0}^{\infty} \boldsymbol{\gamma}^\ell$ from the perspective of graph theory, we form the complete digraph D_n such that the directed edge from vertex i to j is weighted by γ_{ij}. Each term in the sum corresponds to walks of a particular length ℓ. A walk between any two vertices is weighted by the product of the partaking edges. The state variable x_i 'senses' perturbations on any given distinct state variable x_j via the sum total of all the walks in D_n that connect vertex j to vertex i. A given walk can be viewed as amplifying the perturbation exerted at x_j, via the corresponding weight of the walk.

Let us extend the digraph D_n by adding m vertices corresponding to $\mathcal{Z}$~variables and drawing edges between these and the vertices of D_n, where the edge connecting the vertex representing x_i to the vertex representing z_k is labelled κ_{ki}, as shown in Fig. 4.1; the undirected edges induce a bipartite graph structure between the $\mathcal{X}$~ and $\mathcal{Z}$~sets of nodes. The sum over all walks from vertex i in D_n to the vertex representing z_k (following directed edges where applicable) represents the *net propagation* of a perturbation applied at z_k. Thus, if we think of the control system as applying a correction (an adjustment of a given z_k), we can calculate the effect of this correction on any given physiological state variable x_i by taking the collective walks in the graph from the vertex representing z_k to a vertex representing x_i.

Similarly, proceeding in the opposite direction, the sum total of walks from the vertex representing z_k to the one representing x_i (following

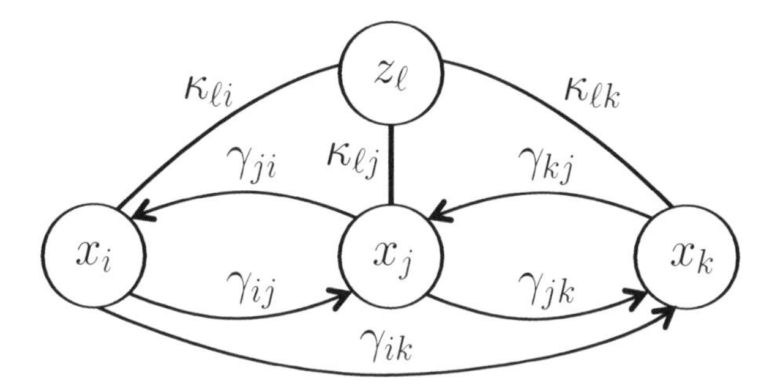

Figure 4.1 **Coupling graph.** The propagation of perturbative effects in the $(\boldsymbol{x}, \boldsymbol{z})$-system can be calculated as a sum-over-walks in a coupling graph, a small portion of which is shown here. The zeroth-order effect is mediated by the κ-type coupling constants between physiological state variables and actuator variables. Higher-order effects arise within the network of $\mathcal{X}$~nodes, encoded by γ-type coupling constants.

directed edges in the *opposite* sense, where applicable) indicates (up to a minus sign) how z_k is to be adjusted in response to the corresponding gradient pressure $\partial L/\partial x_i$. These pressures are summed over all such terms (thus the entire gradient $\nabla_{\boldsymbol{x}} L$ is taken into account) to give the final dynamic adjustment of z_k. This process, for all $z \in \mathcal{Z}$, is represented by the term

$$-\boldsymbol{\kappa} \cdot \left(\sum_{\ell=0}^{\infty} \boldsymbol{\gamma}^{\ell}\right)^{\mathsf{T}} \cdot \nabla_{\boldsymbol{x}} L$$

in Eq. (4.3). By tracing the collected walks from the vertex representing x_i to the vertices representing $z_1, \ldots, z_m$, we glean how a homeostatic pressure arising at x_i translates into pressures at each of the $z \in \mathcal{Z}$.

Let us now return to the working hypothesis that each homeostatic pressure (element of $\nabla_{\boldsymbol{x}} L$) has a 'representative' in a particular neural or neuro-endocrine pathway, or a particular first-messenger (e.g. a particular hormone). This would imply that walks in the coupling graph show how that signal (e.g. hormone) is expected to affect each of the $z \in \mathcal{Z}$. Pleiotropic suites emerge automatically within the coupling graph, which charts how each pressure 'spreads out' over the entire system.

4.2 CONFIGURATIONAL MOTIFS

In the case of transport fluxes or transformations, the coupling coefficients are typically positive in both directions, as the following example illustrates.

Example 12: transport. Consider the following (non-dimensionalised) model for glucose transport φ from an extracellular medium (concentration x_{ext}) to an intracellular medium (concentration x_{int}) [42]:

$$\varphi = \frac{x_{\text{ext}} - x_{\text{int}}}{(x_{\text{ext}} + 1 + \varkappa)(x_{\text{int}} + 1 + \varkappa) - \varkappa^2} \widehat{\varphi} \tag{4.5}$$

where $\varkappa$ and $\widehat{\varphi}$ are positive parameters, $\varkappa$ expressing intrinsic properties of the glucose transporter and $\widehat{\varphi}$ being proportional to its expression level. Here $\widehat{\varphi} \in \mathcal{Z}$. We have $\partial\varphi/\partial x_{\text{ext}} > 0$ and $\partial\varphi/\partial x_{\text{int}} < 0$; however, φ occurs with a plus sign in the dynamics of x_{ext} and with a minus sign in the dynamics of x_{int}, and thus the corresponding terms in the Jacobian matrix (which correspond to the γ-type coupling coefficients) are both positive. On the other hand, the κ-type couplings to a $z \in \mathcal{Z}$ are of opposite signs. This is the general pattern of donor and acceptor control of a flux, although the relative magnitudes will vary according to the particular system under consideration. ❖

Example 13: glucose and fatty acid pools. Consider the digraph depicted in Fig. 4.2. Here the node marked **G** represents the blood plasma glucose concentration ('level') in the mammalian system; **Y** glycogen reserves; **T** triacylglyceride reserves; **F** free (non-esterified) fatty acids in the blood plasma; and **A** the pool of coenzyme A-bound acetyl groups (plus metabolites readily converted into such groups) within the tissues, and which forms the basis for respiratory regeneration of ATP. The $\mathcal{Z}$~nodes have been marked with two-letter codes which are self-explanatory except for the flux marked **A∅** which is the sink flux towards excreted carbon dioxide (source fluxes are not included for the sake of simplicity). Let us assume that blood plasma glucose **G** is represented in L by a conventional convex term, whose minimum corresponds to the physiological optimum. If **G** exceeds this optimum, a positive homeostatic pressure $\partial L/\partial x_{\mathbf{G}}$ arises. Let us apply the walks-in-graph principle with the added stipulation that the shortest path dominates the sign. It is then readily verified that this predicts upward pressures on the fluxes **GA**, **GY**, **YT**, and **A∅**, as well as downward pressures on **YG** and **FA**

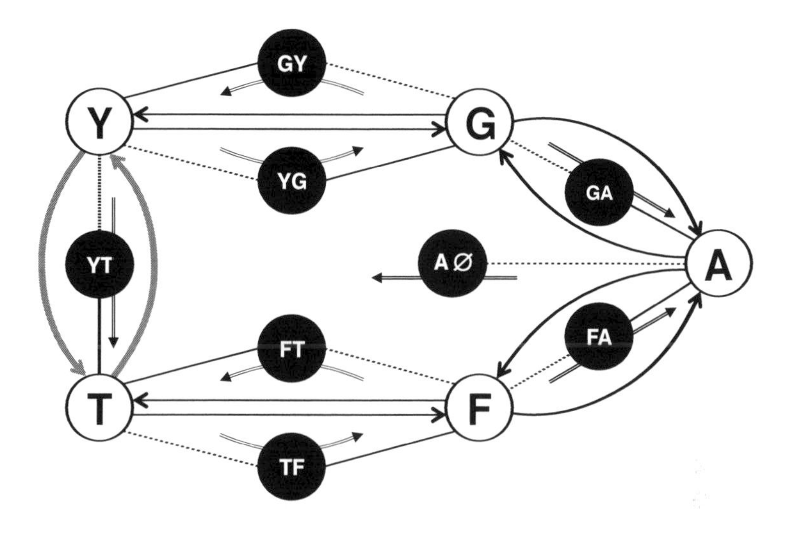

Figure 4.2 **Example of a coupling graph.** Here, $\mathcal{X}$~nodes are depicted in white and $\mathcal{Z}$~nodes are depicted in black. Coupling coefficients of the γ-type are represented by directed edges (arrows) and those of the κ-type by undirected edges. Double-shafted arrows represent the directionality of the transformations and are not formally part of the digraph. Solid lines represent connections of a positive sign, dashed lines negative ones. Labels are explained in the text.

(the latter effect being known as the *glucose-fatty acid cycle* [26]). With **FT** and **TF** we encounter a problem: there are two paths of equal length (one via **Y**-**T**, one via **A**-**F**) that have opposite implications. The relative strength of the couplings determines which prevails. Since **Y** and **T** refer to storage polymers, donor and acceptor control may be supposed to constitute comparatively weak effects, as indicated by grey arrows in Fig. 4.2. Thus we tentatively propose that the **A**-**F** path determines the sense of the effect, i.e. an upward pressure on **FT** and a downward pressure on **TF**.

On the working hypothesis, the list of upward and downward pressures on the $z \in \mathcal{Z}$ exerted by $\partial L/\partial x_{\mathbf{G}} > 0$ translates into the suite of stimulatory and inhibitory effects exerted by this token hormone on the corresponding fluxes and biochemical transformations: the pleiotropic suite associated with the hormone *insulin* [16]. ❖

The foregoing example illustrates both the caveats and the general application of the walks-in-graphs principle. We can understand the phenomenon of hormonal pleiotropy by examining the γ- and κ-type linkages ('causal connections') between the physiological variables, and hence infer the sense (and, where quantitative data are available, the strength) of the myriad effects a given hormone exerts at its various points of action in the organism.

Example 14: nutrient disposition. The previous example can be extended by including interactions with carbohydrate and protein reserves in the muscle; this results in a more complex network as depicted in Fig. 4.3. To eliminate clutter, directed edges have been omitted, but they should be imagined between any given pair of $\mathcal{X}$~nodes that are connected via a $\mathcal{Z}$~node, the latter being represented by smaller filled disks (unlabelled, again to avoid clutter). As in the previous example, all directed edges are of positive sign, as they concern inter-compartmental interactions. Thus, the graph of Fig. 4.2 is a subgraph of that in Fig. 4.3. Although the $\mathcal{Z}$~nodes have not been labelled, their functional significance can readily be inferred from the positive (solid edges) and negative (dashed edges) linkages to the $\mathcal{X}$~nodes. The hypothesis that hormones are tokens for homeostatic pressures is made explicit by the following correspondences:

$$\begin{gathered}
\frac{\partial L}{\partial x_{\mathsf{tag}}} > 0 \leftrightarrow \text{Leptin} \; ; \; \frac{\partial L}{\partial x_{\mathsf{hgl}}} < 0 \leftrightarrow \text{Cortisol} \; ; \\
\frac{\partial L}{\partial x_{\mathsf{glu}}} > 0 \leftrightarrow \text{Insulin} \; ; \; \frac{\partial L}{\partial x_{\mathsf{glu}}} < 0 \leftrightarrow \text{Glucagon} \; ; \\
\frac{\partial L}{\partial x_{\mathsf{naa}}} > 0 \leftrightarrow \text{Growth Hormone} \; ; \\
\frac{\partial L}{\partial x_{\mathsf{naa}}} > 0 \text{ AND } \frac{\partial L}{\partial x_{\mathsf{eaa}}} > 0 \leftrightarrow \text{Insulin-like Growth Factor-1} \, . \qquad (4.6)
\end{gathered}$$

We assign, for each of these hormones, positive or negative effects to each of the $\mathcal{Z}$~nodes, according to the walks-in-graphs recipe under the working assumption that the shortest path will dictate the dominant effect. For instance, if lipid stores (**tag**) exceed their set-point, the path analysis predicts a *stimulatory* pressure on the glucose utilisation flux **glc - - • — aca** (via positive donor-acceptor γ-type couplings between **tag** and **hgl**, as well as **hgl** and **glc**, a negative link from **glc**, and the sign reversal from the general formula) and this would suggest that

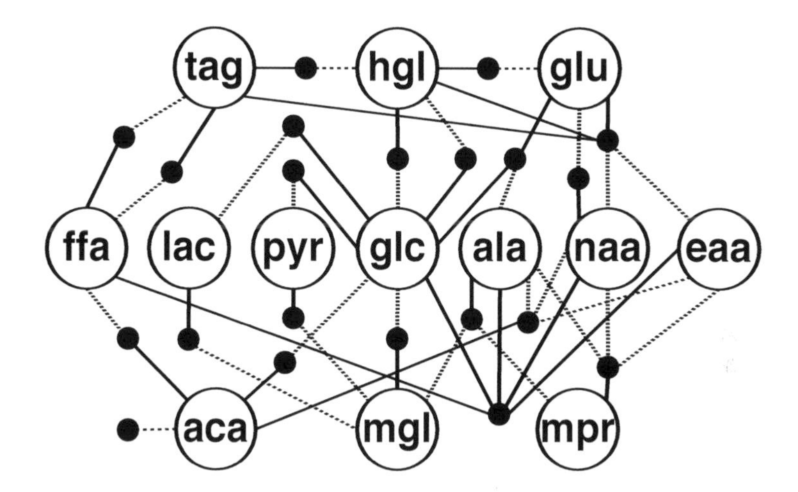

Figure 4.3 **Coupling graph for the mammalian nutrient disposition system.** Solid edges represent positive κ-type couplings; dashed edges represent negative κ-type couplings. The $\mathcal{Z}\sim$nodes are represented as black disks, whereas open disks with three-letter labels are $\mathcal{X}\sim$nodes, as follows: **tag**: triacylglyceride stores (lipid reserves); **hgl**: hepatic glycogen stores (carbohydrate reserves); **glu**: hepatic glutamate pool; **ffa**: free fatty acids in blood plasma; **lac**: lactate in blood plasma; **pyr**: pyruvate in blood plasma; **glc**: glucose in blood plasma; **ala**: alanine in blood plasma; **naa**: non-essential amino acids (other than alanine) in blood plasma; **eaa**: essential amino acids in blood plasma; **aca**: acetyl-CoA (or equivalent metabolite) pool in all tissues; **mgl**: muscle glycogen stores (carbohydrate reserves); **mgl**: muscle protein stores (protein reserves). Positive γ-type couplings (not shown) are assumed to exist between every pair of $\mathcal{X}\sim$nodes that is connected by a $\mathcal{Z}\sim$node.

administering leptin to an experimental animal should stimulate glucose metabolism, as experimental studies confirm [40].

Similarly, one would predict that both leptin and insulin have a negative effect on food intake, which in the graph in Fig. 4.3 is represented by the • node with positive (solid line) connections to **ffa**, **glc**, **ala**, **naa**, and **eaa**. In fact, such inhibitory effects exist and are mediated by neuroendocrine pathways in the hypothalamus [35, 98].

Rather subtle stoichiometric requirements are hidden in the • node with a positive (solid line) connection to **mpr**, as well as negative (dashed line) connections to **ala**, **naa**, and **eaa**. The donors of this flux are the free amino acids in the blood plasma, and the manner in which the flux co-depends on the concentrations of all of these is complicated. The crux of the matter is that the relative proportions of uptake are subject to comparatively loose constraints, at least to the extent that uptake can be made to replenish the acceptor cells' pool of amino groups (which is managed through transamination reactions [26, 58]) and thus counteract proteolysis and to feed the synthesis of key proteins. On the other hand, proliferative protein synthesis is associated with much stricter stoichiometric constraints [26, 58], the essential amino acids (**eaa**) representing the major bottle neck, since whereas they can be converted into non-essential amino acids (**naa**), the converse is not true [26, 77]. This stoichiometric harmonisation is achieved by associating separate signals with surplus pressures from **naa** (predominantly via arginine [2, 55]) and **eaa** as indicated in Eq. (14). Here, the AND-function is performed by the liver cells which secrete IGF-1 in response to GH stimulation, with a responsiveness that is sharply potentiated by the presence of **eaa** in the blood plasma [97].

Of course, the graph-based approach does not capture every single aspect of endocrine regulation of nutrient disposition and energetics in mammals[†]. Nonetheless, this approach allows us to connect endocrinological and physiological insights, which lends a modicum of credence to the notion that the pleiotropic effects of any given hormone (or first-messenger, more generally) can be understood in terms of the 'flux logic' of the underlying physiological system. ❖

[†]To mention one fascinating example: the arcuate nucleus of the rat hypothalamus contains neurons whose activity levels correlate with the blood plasma glucose level (**glc**), but with *strikingly different time constants* across an array of subtypes [66]. This implies that the brain is capable, in principle, of responding to the rate of change of **glc** over several time scales (e.g., seconds, minutes, hours). Classic control engineering rests on three pillars, PID, where P = proportional, I = integrating, and D = differentiating or derivative-based [39]. We encountered P and I in Chapter 2, but have had little to say about D-control, which may well be underpinned by these remarkable sensory neurons.

The graphical representations of mammalian nutrient disposition in Figs 4.2 and 4.3 correspond to high-dimensional dynamical systems: each node corresponds to a state variable, inducing a first-order ODE. Analysis of the dynamical flow of such systems may become quite involved. However, we can gain some insight into the overall qualitative characteristics of this dynamical flow by breaking these diagrams down into *design motifs*. Two such motifs are discussed in the following examples.

Example 15: hub-and-reservoir. The first motif is the hub-and-reservoir configuration shown in Fig. 4.4. All the nodes in the middle row of Fig. 4.3 can be regarded as 'hubs' and the nodes labelled **tag**, **hgl**, **mgl**, and **mpr** in the same figure as 'reservoirs' which suggests that this motif is prevalent. As shown in Fig. 4.4, it essentially consists of two pools connected by $\mathcal{Z}$~variables which physically correspond to fluxes or conversions (or both). One of the pools receives additional fluxes, which in the wider model would correspond to $\mathcal{Z}$~variables pertaining to that wider model. However, for the local context these fluxes are treated as a lumped input u. As the latter varies in time, the total amount of matter represented by x_1 and x_2 together does not stay constant. This implies that strict (quasi-optimal) homeostasis for both cannot be achieved. It may well be the case that regulatory loops in the wider system work to keep $|u|$ as small as possible, but our focus here is on the 'local' system consisting just of x_1, x_2, z_{12}, and z_{21}.

If we scale x_1 and x_2 with respect to their optimal values and expand to leading order, assuming no interaction term between x_1 and x_2, we

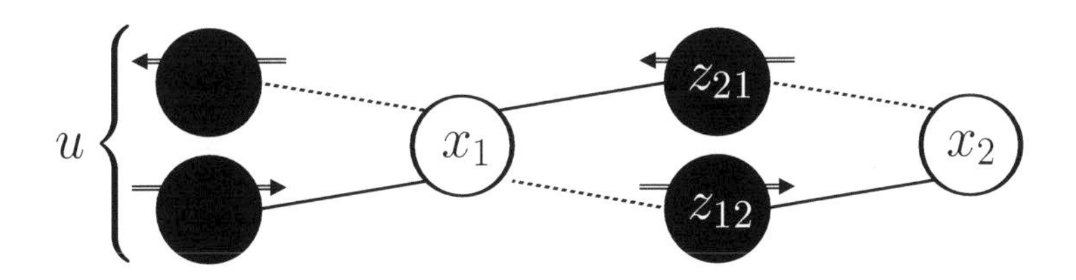

Figure 4.4 **Hub-and-reservoir configuration.** Solid edges represent positive κ-type couplings; dashed edges represent negative κ-type couplings. The $\mathcal{Z}$~nodes are represented as black disks and $\mathcal{X}$~nodes are depicted as open disks with three-letter labels.

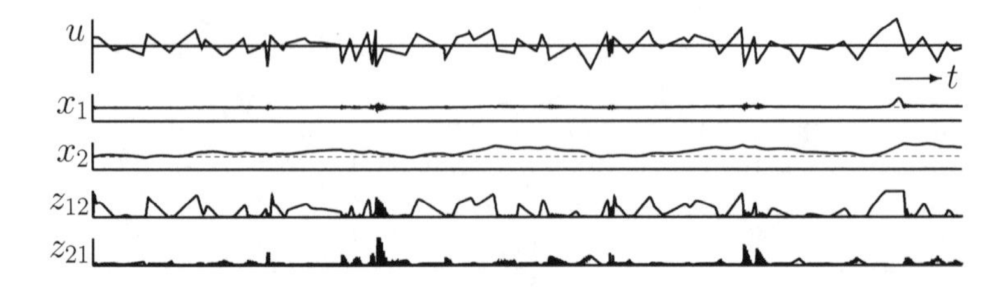

Figure 4.5 **Hub-and-reservoir dynamics.** Time traces of the variables introduced in the previous figure. The top trace is the input function $u(t)$ and the next four traces show the behaviour of the state $\{x_1(t), x_2(t), z_{12}(t), z_{21}(t)\}$ under this input.

arrive at the following form:

$$L(x_1, x_2, z_1, z_2) = (x_1 - 1)^2 + \rho(x_2 - 1)^2 + \text{term(s) corresponding to } z_1 \text{ and } z_2 \quad (4.7)$$

with $\rho > 0$ such that for $\rho \ll 1$ homeostasis would be achievable for x_1, at the cost of fluctuations in x_2, whereas when $\rho \gg 1$, the reverse obtains. The scaling with respect to the optima induces a conversion factor; that is, the scaled balance equations take on the following form:

$$\begin{aligned} \dot{x}_1 &= u - z_{12} + \gamma z_{21} \\ \dot{x}_2 &= u + z_{12}/\gamma - z_{21} \end{aligned} \quad (4.8)$$

such that with $\gamma > 1$, x_1 is the hub and x_2 is the reservoir, and *vice versa* for $0 < \gamma < 1$. In the case $\gamma = 1$, there is no clear-cut division and the hub/reservoir distinction would be moot. However, ρ and γ work together: for instance, for $\rho \ll 1$, the fluctuations predominantly experienced by x_2 are lessened in *relative* terms if $\gamma > 1$. A similar argument applies to the conditions $\rho \gg 1$, $\gamma < 1$.

We see that either x_1 is the hub and x_2 is the reservoir, or the other way around. This prompts us to ask which considerations break the symmetry. Let us entertain the following proposition (cf. Section 2.1): the pool that communicates with the wider system (via u) should be the one that experiences strict homeostasis; in Fig. 4.4, this pool is x_1 (the labelling is arbitrary). Fluxes that are supplied by this pool can then be efficiently regulated by up- or down-regulating the machinery (enzymes, transporters, etc.) that mediates these fluxes, since their substrate species is maintained at a near-constant level; in this manner the usage traffic is shielded off from the fluctuations at the level of supply to this

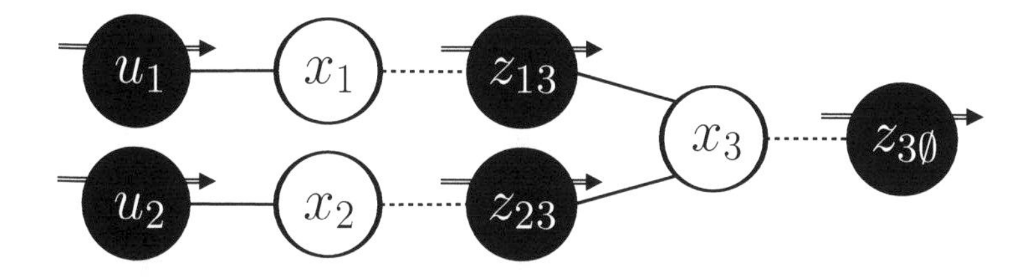

Figure 4.6 **Alternate sources configuration.** Solid edges represent positive κ-type couplings; dashed edges represent negative κ-type couplings. The $\mathcal{Z}\sim$nodes are represented as black disks and $\mathcal{X}\sim$nodes as open disks with three-letter labels.

substrate pool. An additional consideration is that the hub substrate is typically an osmotically active species of small-molecular weight. Therefore, controlling its fluctuations becomes part of the organism's hydromineral regulation.

In Eq. (4.7), the term corresponding to $\mathcal{Z}\sim$variables was left unspecified. A control surface as depicted in the right-most panel of Fig. 3.2 would seem plausible since this forbids z_{12} and z_{21} to be both high at the same time, since this would result in an idle cycle. Combining Eqs. (4.4) and (4.8), we obtain dynamics that can be evaluated numerically, as shown in Fig. 4.5. Here $\rho = 10$ and $\gamma = 0.1$, making x_1 the tightly controlled hub. Thus x_1 tends to stay close to the optimal value (dashed line) whereas x_2 tends to range more widely. We have $z_{12} \lesssim \max\{0, u\}$ and $z_{21} \lesssim \max\{0, -u\}$.

As formulated here, x_2 is allowed to become negative, which is not realistic from a biological point of view. An Anschlag term should be added to Eq. (4.7) to ensure that $x_2 \geq 0$. As x_2 approaches zero, the Anschlag gradient overwhelms the gradient associated with x_1. As a result, homeostasis of x_1 is broken or 'frustrated.' ❖

We have discussed the hub-and-reservoir configuration in terms of exchange fluxes between tissues and organs in a multicellular organism, but similar considerations apply at the sub-cellular level. Catabolic and anabolic pathways both converge on, and depart from, a small number of interconnected pools of metabolites with low molecular weight, and temporary surpluses are absorbed by polymeric reserve compounds, which can occupy substantial portions of the total intracellular volume in bacteria [50].

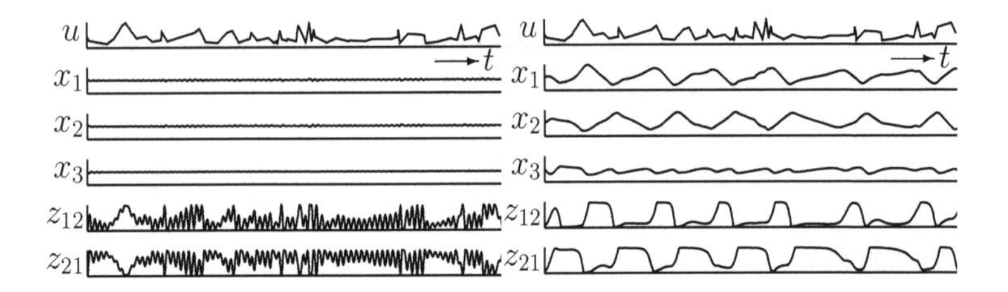

Figure 4.7 **Dynamics of the alternate sources configuration.** Time traces of the variables introduced in the previous figure. The top trace is the input function $u(t)$ and the next five traces show the behaviour of the state $\{x_1(t), x_2(t), x_3(t), z_{12}(t), z_{21}(t)\}$ under this input. The left shows a 'reactive' control regime, the right a 'slack' one, both in response to the same irregular input $u(t)$.

The hub-and-reservoir motif occurs twice in Fig. 4.2, with hub **G** interacting with reservoir **Y** and with hub **F** interacting with reservoir **T**. This graph shows further features, such as an alternate supply configuration involving **G**, **F**, and **A**. The latter is explored in the next example.

Example 16: alternate sources. The configuration is schematically represented in Fig. 4.6. We assume that the demand flux $z_{3\emptyset}$ is constant and that the inputs u_1 and u_2 are such that $u_1 + u_2$ is constant. Appropriate control surfaces are as in the middle and left-most panels of Fig. 3.2. Examples of the dynamics of this configuration are shown in Fig. 4.7; these are merely two instances of the variety that can be created by varying the parameter values, chosen here to demonstrate the contrast between reactive ('jittery,' 'aggressive') z with near-perfect control of $\boldsymbol{x}$ (left-hand panel of Fig. 4.7), to be contrasted to more smooth ('inert') behaviour of the z, but with more pronounced variation in $\boldsymbol{x}$ (right-hand panel of Fig. 4.7; the irregular input $u(t)$ is identical in this comparison). In these examples, L contains terms of the form $(x_i - 1)^2$, $i = 1,2,3$, to L. By weighing the term $(1 - x_3)^2$ with a factor greater than 1, smoothness in z_1, z_2 can be paired with near-perfect regulation of x_3, at the cost of marked fluctuations in x_1 and x_2. ❖

CHAPTER 5

Differential inclusions

Let us consider the reduced dynamics $\dot{\boldsymbol{x}} = \widetilde{\boldsymbol{f}}(\boldsymbol{x},\boldsymbol{u})$, obtained from the full dynamics $\dot{\boldsymbol{x}} = \boldsymbol{f}(\boldsymbol{x},z,\boldsymbol{u})$ by setting $z = \overline{\boldsymbol{g}}(\boldsymbol{x},\boldsymbol{u})$. The conditions that $\overline{\boldsymbol{g}}$ has to satisfy under the QSC approximation are specified by Eqs. (2.9) and (3.8). If we fix the input $\boldsymbol{u}(t) \equiv \overline{\boldsymbol{u}}$, we obtain autonomous dynamics $\widetilde{\boldsymbol{f}}$ on $\mathcal{X}$. We can study the flow associated with $\widetilde{\boldsymbol{f}}$ by first classifying, at each point $x \in \mathcal{X}$, the elements of $z = \overline{\boldsymbol{g}}(\boldsymbol{x},\overline{\boldsymbol{u}})$ according to whether they are singular or non-singular (i.e., 'stopped'), and if the latter, at which value they are stopped (e.g., a lower or an upper boundary of an interval of allowed values). If there are m $\mathcal{Z}$~variables that all have a simple range restriction, as in Eq. (2.18), then for each z_j we have either $z_j = z_{j,\min}$, $z_{j,\min} < z_j < z_{j,\max}$, or $z_j = z_{j,\max}$. On this classification of the elements of z $(= \overline{\boldsymbol{g}}(\boldsymbol{x},\overline{\boldsymbol{u}}))$, we obtain a partition of $\mathcal{X}$ into 3^m subsets. The essential qualitative characteristics of the system are well-represented as a series of transitions between these 3^m regions (i.e. the *symbolic dynamics* of x relative to this partition).

In the previous chapters, the 'flat' Anschlag played a special role. We recall that it has the property that L does not vary with z_j whenever $z_j \in (z_{j,\min}, z_{j,\max})$. In this case, the dynamics can be treated as a *differential inclusion* (DI), in particular a piecewise-smooth dynamical system; see e.g. refs. 6, 20, and 81 for introductions to the mathematical foundations. In particular, to find in the limit the DI corresponding to the dynamics when control is non-extremal, we may depart from the Anschlag as defined by Eq. (3.2), and then consider the limit $\epsilon \to 0$. If the dynamics on $\mathcal{X}$ is sufficiently smooth (in a sense which need not detain us here, but see ref. 20), $\mathcal{X}$ is partitioned by a constellation of sliding regions. The result is a global, geometric view of the behaviour of the system, as claimed by posit 5 in Section 1.2.

In Section 3.4 we addressed the generic problems associated with attempting to evaluate $\overline{g}$ directly, and we introduced Eq. (3.21), which lifts the dynamics from $\mathcal{X}$ to a phase space $\mathcal{X} \times \mathcal{Z}$. The DI approximation brings us back down to $\mathcal{X}$. Taking the limit $\mu_j \to \infty \; \forall j$ in Eq. (3.21), which creates time-scale separation between $\mathcal{X}$~ and $\mathcal{Z}$~variables, along with the aforementioned 'hard-stop' limit $\epsilon \to 0$, in Eq. (3.2), we move from a smooth dynamical system on $\mathcal{X} \times \mathcal{Z}$ to a DI on $\mathcal{X}$. One may query the biological plausibility of these limits — the chattering behaviour and infinite rates of change of the $\mathcal{Z}$~variables hardly strike one as physiologically realistic. Even so, the DI on the lower-dimensional space $\mathcal{X}$ is worth careful consideration, inasmuch as it gives a qualitative insight into the more intricate dynamical flow on $\mathcal{X} \times \mathcal{Z}$.

Example 17: temperature control (continued). In Example 11, we considered Eq. (3.23) describing a system of ODEs for x and z. As long as the input $u(t)$ is confined to the regulatory range, the actuator variable z can be adjusted to bring and keep the temperature (the physiological state) x close to the optimal value 1. As u fluctuates within the regulatory range, z moves within the interval $[z_{\min}, z_{\max}]$ whilst x ranges over a neighbourhood of 1 that can be made as small as we please by taking μ sufficiently large and η sufficiently small. On the other hand, when u travels outside of the regulatory range, z is stopped at one of the Anschläge and x is allowed to drift away from 1.

Taken together, these observations suggest an effective reduction of the number of dynamical degrees of freedom from 2 to 1. Although the system has two dynamical degrees of freedom, one of them is effectively frozen at almost every moment in time: either x is at (or very near) the value 1 while z is being adjusted within the regulatory range, or else z is stopped at $z_{\min}$ or $z_{\max}$, while x wanders away from the optimal value. Let us consider the following differential inclusion (DI):

$$\dot{x} \begin{cases} = u - x + z_{\max} & \text{for } x < 1 \\ \in [u - x + z_{\min}, u - x + z_{\max}] & \text{for } x = 1 \\ = u - x + z_{\min} & \text{for } x > 1 \end{cases} . \tag{5.1}$$

The model we have been studying converges (in the sense of point-wise convergence of the system trajectories) to this DI under the limits $\epsilon \to 0$ (for the Anschläge), $\mu \to \infty$ (in Eq. (3.23)), $\eta \to 0$ (in Eq. (3.22)). ❖

According to Eq. (5.1), the flow on $\mathcal{X}$ is almost everywhere to be treated as autonomous relative to the actuator variable, which is 'stopped' at an extreme allowed value. The exception is formed by an affine subspace, the *sliding manifold*, in which $\dot{x}$ is set to a value taken from a set of available values, such that x remains in this affine subspace. The z-value corresponding to this choice is called a *singular value.*

In Example 17, the sliding manifold is extremely simple in structure (consisting of no more than the point $x = 1$) and only the value $\dot{x} = 0$ is available. The geometry of the affine subspace and the possible motions on this subspace become more involved when $\mathcal{X} = \mathbb{R}^n$ for $n > 1$, as we shall presently see. Throughout we will assume the external input $\boldsymbol{u}(t)$ to be fixed at some constant value $\overline{\boldsymbol{u}}$. This simplifies the arguments considerably without materially affecting the main conclusions.

5.1 THE DISCONTINUITY HYPERPLANES DEFINING THE REGULATORY RANGE

We consider dynamics as in Eq. (4.4) with the following expression for L:

$$L(\boldsymbol{x}, z) = \frac{1}{2}\sum_{i=1}^{n}(x_i - 1)^2 + \sum_{j=1}^{m}\left(\frac{\eta}{4}\left(\frac{2z_j - z_{j,\max} - z_{j,\min}}{z_{j,\max} - z_{j,\min}}\right)^2 + \sigma_\epsilon(z_j)\right) \quad (5.2)$$

with $\eta > 0$; here σ_ϵ denotes Anschlag terms as in Eq. (3.2), with $\epsilon < (z_{j,\max} - z_{j,\min})/2$. By definition, the z_j null isocline is the locus of $\dot{z}_j = 0$; for $z_{j,\min} + \epsilon \leq z \leq z_{j,\max} - \epsilon$ this is the hyperplane defined by

$$z_j = \frac{z_{j,\max} - z_{j,\min}}{2}\left(1 - \eta^{-1}\sum_{i=1}^{n}\frac{\kappa_{ji}}{\mu_j}(x_i - 1)\right). \quad (5.3)$$

Of special interest are three regions of $\mathbb{R}^n$ which are defined in reference to the z_j null isocline: (i) $\mathcal{L}_j^{[+\epsilon]} \in \mathbb{R}^n$ defined as the locus of

$$z_{j,\min} + \epsilon = \frac{z_{j,\max} - z_{j,\min}}{2}\left(1 - \eta^{-1}\sum_{i=1}^{n}\frac{\kappa_{ji}}{\mu_j}(x_i - 1)\right); \quad (5.4)$$

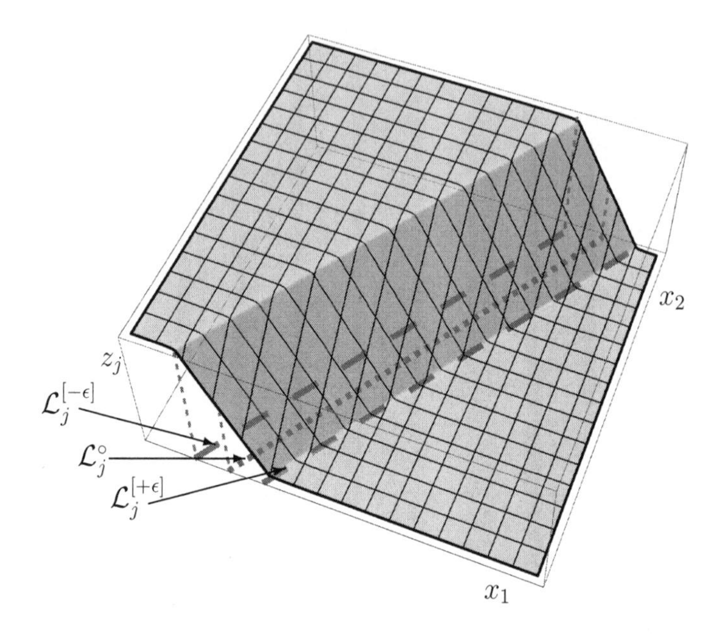

Figure 5.1 **Loci in $\boldsymbol{x}$-space.** The surface is a graph of the z_j-null isocline as a function of $\boldsymbol{x} = (x_1, x_2)$. The dashed lines in the (x_1, x_2)-plane are the loci $\mathcal{L}_j^{[-\epsilon]}$, $\mathcal{L}_j^{\circ}$, and $\mathcal{L}_j^{[+\epsilon]}$, as indicated. The lines $\mathcal{L}_j^{[-\epsilon]}$ and $\mathcal{L}_j^{[+\epsilon]}$ correspond to discontinuities in the null isocline. Steepening up the central portion of the surface, we obtain a convergence of $\mathcal{L}_j^{[-\epsilon]}$, $\mathcal{L}_j^{\circ}$, and $\mathcal{L}_j^{[+\epsilon]}$.

(ii) $\mathcal{L}_j^{\circ} \in \mathbb{R}^n$, the locus of $\sum_{i=1}^{n} \kappa_{ji}(x_i - 1) = 0$; and (iii) $\mathcal{L}_j^{[-\epsilon]} \in \mathbb{R}^n$, the locus of

$$z_{j,\max} - \epsilon = \frac{z_{j,\max} - z_{j,\min}}{2} \left(1 - \eta^{-1} \sum_{i=1}^{n} \frac{\kappa_{ji}}{\mu_j} (x_i - 1) \right). \tag{5.5}$$

These definitions are illustrated for $n = 2$ in Fig. 5.1. For $n = 2$ the three loci are 1-dimensional; in general they are hyperplanes of dimension $n - 1$. The separation between these three hyperplanes is of order η and both $\mathcal{L}_j^{[+\epsilon]}$ and $\mathcal{L}_j^{[-\epsilon]}$ converge to $\mathcal{L}_j^{\circ}$ as $\eta \to 0$.

To characterise the direction of flow in the system, fix a point $\boldsymbol{x}^{\circ} \in \mathcal{L}_j^{\circ}$ and let $\boldsymbol{x}^{\circ+\epsilon} \in \mathcal{L}_j^{[+\epsilon]}$ such that $\|\boldsymbol{x}^{\circ} - \boldsymbol{x}^{\circ+\epsilon}\| \leq \|\boldsymbol{x}^{\circ} - \boldsymbol{x}'\|$ for all $\boldsymbol{x}' \in \mathcal{L}_j^{[+\epsilon]}$ (i.e., $\boldsymbol{x}^{\circ+\epsilon}$ is the nearest neighbour of $\boldsymbol{x}^{\circ}$ in $\mathcal{L}_j^{[+\epsilon]}$). The difference $\boldsymbol{x}^{\circ} - \boldsymbol{x}^{\circ+\epsilon}$ is the vector pointing from $\boldsymbol{x}^{\circ+\epsilon}$ to $\boldsymbol{x}^{\circ}$, provided that the latter are regarded as position vectors relative to the origin. Let $\mathcal{I}^{[+\epsilon]}(\boldsymbol{x}^{\circ}, z)$ denote the scalar product $\boldsymbol{f}(\boldsymbol{x}^{\circ+\epsilon}, z)\cdot(\boldsymbol{x}^{\circ}-\boldsymbol{x}^{\circ+\epsilon})$. The dynamic flow $\boldsymbol{f}$ locally points into the region bounded by $\mathcal{L}_j^{[+\epsilon]}$ and $\mathcal{L}_j^{[-\epsilon]}$ precisely when $\mathcal{I}^{[+\epsilon]}(\boldsymbol{x}^{\circ}) > 0$. We similarly define the point $\boldsymbol{x}^{\circ-\epsilon} \in \mathcal{L}_j^{[-\epsilon]}$ with its associated scalar projection $\mathcal{I}^{[-\epsilon]}(\boldsymbol{x}^{\circ}, z)$. Let us say that $\boldsymbol{x}^{\circ} \in \mathcal{L}_j^{\circ}$ is (i) *weakly locally attracting* if and only if both $\mathcal{I}^{[+\epsilon]}(\boldsymbol{x}^{\circ}, z) > 0$ and $\mathcal{I}^{[-\epsilon]}(\boldsymbol{x}^{\circ}, z) > 0$; (ii) *weakly locally repelling* if and only if both $\mathcal{I}^{[+\epsilon]}(\boldsymbol{x}^{\circ}, z) < 0$ and $\mathcal{I}^{[-\epsilon]}(\boldsymbol{x}^{\circ}, z) < 0$; and (iii) *weakly locally transversal* if and only if $\mathcal{I}^{[+\epsilon]}(\boldsymbol{x}^{\circ}, z)\mathcal{I}^{[-\epsilon]}(\boldsymbol{x}^{\circ}, z) < 0$.

These weak characteristics do not, as yet, permit us to draw firm conclusions about the dynamic flow in the vicinity of $\mathcal{L}_j^{\circ}$ — further regularity assumptions on $\boldsymbol{f}$ are required. Let us first restrict our attention to regions (i.e., compact subsets) of $\mathcal{L}_j^{\circ}$ such that all points in this subset share a common index signature $\left\{\mathcal{I}^{[+\epsilon]}(\boldsymbol{x}^{\circ}, z), \mathcal{I}^{[-\epsilon]}(\boldsymbol{x}^{\circ}, z)\right\}$. We consider the flow in the neighbourhood of any point $\boldsymbol{x}^{\circ}$ internal to such a subset. Finally, we require that every $\mathcal{Z}\sim$variable is no further than $\widetilde{\epsilon}$ removed from either its minimum or its maximum value:

$$\min\{z_\ell - z_{\ell,\min}, z_{\ell,\max} - z_\ell\} \leq \widetilde{\epsilon}$$

with $\epsilon < \widetilde{\epsilon} < (z_{\ell,\max} - z_{\ell,\min})/2$ for $\ell \in \{1,\ldots,m\}$. Furthermore, we assume that the index signature is invariant as these $\mathcal{Z}\sim$variables vary independently within their allowed ranges, which are intervals of width $\widetilde{\epsilon}$ adjacent to either $z_{\ell,\min}$ or $z_{\ell,\max}$. Under these assumptions, the dependence on z can be unambiguously dropped from the index signature, and it can be assigned to a region $\mathcal{R} \subset \mathcal{L}_j^{\circ}$ itself, thus: $\left\{\mathcal{I}^{[+\epsilon]}(\mathcal{R}), \mathcal{I}^{[-\epsilon]}(\mathcal{R})\right\}$. If the latter is $\{+,+\}$ and the assumptions on the flow hold good, we can say that $\mathcal{R}$ is *locally attracting*, and correspondingly for the other sign signatures.

The reason for allowing a wider margin $\widetilde{\epsilon}$ is that z_j will not quite have attained $z_{j,\min}+\epsilon$ when $\boldsymbol{x}$ traverses $\mathcal{L}_j^{[+\epsilon]}$, or $z_{j,\max}-\epsilon$ when $\boldsymbol{x}$ traverses $\mathcal{L}_j^{[-\epsilon]}$. The 'maximum travel time' from the other extreme of z_j's range has to be taken into account. This travel time scales as μ_j^{-1} and accordingly vanishes as μ_j is allowed to increase without bound. In the limit $\mu_j \to \infty$ a time-scale separation arises between the dynamics of $\boldsymbol{x}$ and z_j, with the latter virtually moving along its null-isocline, as a quasi-static function of $\boldsymbol{x}$.

Even if $\mathcal{R} \subset \mathcal{L}_j^{\circ}$ is locally attracting, not much can be said about the flow $\boldsymbol{f}$ once it is confined between the boundaries $\mathcal{L}_j^{[+\epsilon]}$ and $\mathcal{L}_j^{[-\epsilon]}$. The flow may be contained in the sense that for $t \to \infty$ (or in any case until external conditions cause $\boldsymbol{u}$ to change) there will always be an $\boldsymbol{x}^{\circ}(t) \in \mathcal{R}$

such that $\|\boldsymbol{x}(t) - \boldsymbol{x}^\circ(t)\| < \delta_{\mathcal{R}}$, where $\delta_{\mathcal{R}}$ is the lowest upper bound such that every $\boldsymbol{x}^\circ \in \mathcal{R}$ has points in both $\mathcal{L}_j^{[+\epsilon]}$ and $\mathcal{L}_j^{[-\epsilon]}$ contained within the $\delta_{\mathcal{R}}$-ball centered on $\boldsymbol{x}^\circ$. Yet, even in this case there is no guarantee that all state variables converge to the physiological optimum (i.e., $x_i \to 1$ for all i); in other words, the system may remain *homeostatically frustrated* as no feasible combination of values for the z_j is such that the condition $x_i = 1\ \forall i$ can be satisfied.

If the flow is not contained in the above sense, $\boldsymbol{x}(t)$ exits the volume of phase space consisting of all points within $\delta_{\mathcal{R}}$-distance of at least one point in $\mathcal{R}$. For instance, the trajectory may enter a neighbouring region $\mathcal{R}'$ whose sign signature is $\{-,-\}$; more precisely, $\boldsymbol{x}(t)$ enters the subvolume of phase space consisting of points within $\delta_{\mathcal{R}'}$-distance of at least one point in $\mathcal{R}'$. Under the stated assumptions, the trajectory exits this subvolume. As soon as this happens, z_j will relax (at rate μ_j) towards a value no further than ϵ away from either $z_{j,\min}$ or $z_{j,\max}$. In other words, $z_j(t)$ will be *stopped* at an Anschlag.

5.2 PASSING TO THE DIFFERENTIAL INCLUSION

Let us now consider what happens when we project down to the reduced dynamics on $\mathcal{X}$. The subtlety resides in the fact that $\overline{\boldsymbol{g}}$ is a multifunction and $\widetilde{\boldsymbol{f}}$ becomes a set of cardinality greater than 1, which means the dynamics becomes a *differential inclusion* [6, 20, 81]. We pass, for all $j \in \{1,\ldots,m\}$, to the limits $\mu_j \to \infty$, $\kappa_{ji} \to \infty$ such that κ_{ji}/μ_j is fixed for every (i,j), $\eta_j \to 0$ and $\widetilde{\epsilon} \to 0$. In these limits, $\delta_{\mathcal{R}} \to 0$ for any region $\mathcal{R} \subset \mathcal{L}_j^\circ$ of interest; that is, the corresponding regions of the boundaries $\mathcal{L}_j^{[+\epsilon]}$ and $\mathcal{L}_j^{[-\epsilon]}$ both converge onto $\mathcal{R}$.

Example 18: transformation of the phase portrait as $\eta_j \to 0$, $\widetilde{\epsilon} \to 0$, for $n = m = 1$. The convergence onto the fairly simple differential inclusion for the case $n = m = 1$ is illustrated in Fig. 5.2. Here the loci $\mathcal{L}^{[+\epsilon]}$, $\mathcal{L}^\circ$, and $\mathcal{L}^{[-\epsilon]}$ are singleton sets, corresponding to three points on the x-axis; in the double limit, the three converge to $x = \widehat{x} = 1$. Since z takes on its extreme values to the left and to the right of $x = 1$, this is a *discontinuity point* for z; it is the lowest-dimensional version of the *discontinuity hyperplanes* for $\mathcal{Z}$~variables: these are regions where these variables can take on their singular (i.e., non-extreme) values. ❖

For every point $\boldsymbol{x}^\circ \in \mathcal{R}$, we have $\lim_{\eta \to 0} \boldsymbol{x}^{\circ+\epsilon} = \lim_{\eta \to 0} \boldsymbol{x}^{\circ-\epsilon} = \boldsymbol{x}^\circ$, but this does not imply convergence of the scalar indices $\mathcal{I}^{[+\epsilon]}(\boldsymbol{x}^\circ, \boldsymbol{z})$

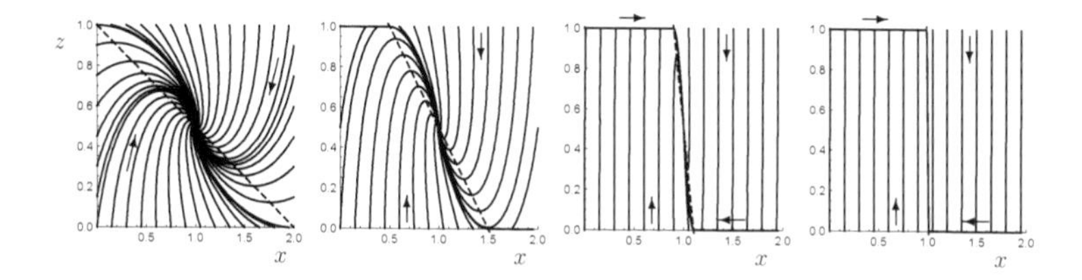

Figure 5.2 **Metamorphosis: smooth flow to differential inclusion (phase portraits).** The phase flow shown is for the system (x,z) with dynamics given by Eqs. (2.4) and (5.2) for $n = m = 1$, with $\mu = \kappa = \eta^{-2}$, $\epsilon = 0.001$, and $u = 0.5$. From left to right, η has the following values: 1, 0.5, 0.1, and 0.01. The general direction of the phase flow is indicated by arrows. In the limit, z relaxes quasi-instantaneously to either $z_{\min} = 0$ or $z_{\max} = 1$ whenever $x \neq 1$; the point $x = 1$ is here the locus $\mathcal{L}^{\circ}$ on which z attains its singular value.

and $\mathcal{I}^{[-\epsilon]}(\boldsymbol{x}^{\circ}, z)$, since the former is determined by the local flow with z_j set to $z^{+}_{j,\min}$ whereas the latter is determined by the local flow with z_j set to $z^{-}_{j,\max}$. In other words, $\mathcal{R}$ retains its index sign signature, but now in the setting of a differential inclusion, in which either one of the signatures $\{+,+\}$ and $\{-,-\}$ is said to characterise $\mathcal{R}$ as a *sliding region*; specifically, the motion of $\boldsymbol{x}$ through a sliding region is *attracting* when the signature is $\{+,+\}$ and *repelling* otherwise. Since the former is of primary interest, 'sliding motion' is usually taken to mean 'attracting sliding motion' [20]. The mixed signatures $\{+,-\}$ and $\{-,+\}$ characterise $\mathcal{R}$ as a *transversal hyperplane*.

Example 19: sliding motions for $n = m = 2$. Various possible cases are shown in Fig. 5.3. The pair (z_1, z_2) gives rise to two lines $(\mathcal{L}^{\circ}_1, \mathcal{L}^{\circ}_2)$ along which sliding motion may occur. At points (x_1, x_2) bounded away from these lines, the z_j are *stopped* at their extreme values. On these lines, and in a neighbourhood whose width is controlled by $\widetilde{\epsilon}$, the z_j assume intermediate values that maintain the phase point on $\mathcal{L}^{\circ}_j$ as it slides along such a line. Such non-extremal z-values will be referred to as *singular*. Thus, in the limit $\widetilde{\epsilon} \to 0$, the z_j are generically singular whenever the phase point (x_1, x_2) is confined to the lines $(\mathcal{L}^{\circ}_1, \mathcal{L}^{\circ}_2)$ and non-singular (stopped) when (x_1, x_2) is not on these lines. In the regions bounded by these lines we have two-dimensional dynamics,

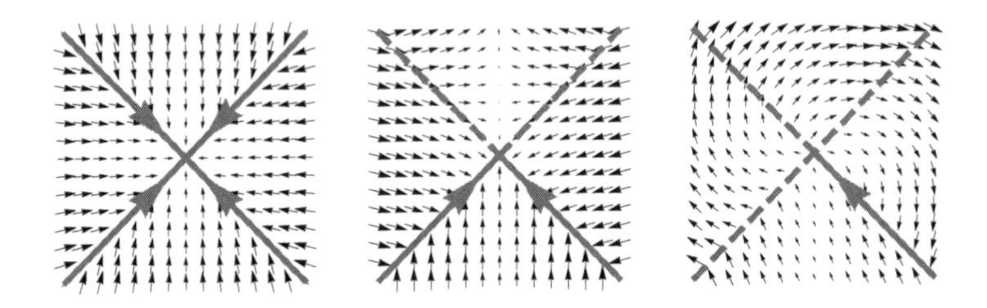

Figure 5.3 **Examples of the signature structure in the neighbourhood of the homeostatic equilibrium point.** All three panels depict the vector field of f for the case $n = m = 2$, in the neighbourhood of the homeostatic equilibrium point at the centre. The plane is the (x_1, x_2)-plane. Lines are the loci $\mathcal{L}_1^\circ$ and $\mathcal{L}_2^\circ$, drawn solid when attracting, dashed when transversal.

since the z_j are fixed at extreme values, and on the lines we have one-dimensional *sliding mode* dynamics (whereas the full dynamics has $n + m = 4$ degrees of freedom). The corresponding flow field is depicted by arrows in Fig. 5.3. The left-most panel depicts a generic situation, where the sliding motion along the discontinuity lines (singular lines, having signature $\{+,+\}$) is such that the phase point (x_1, x_1) is eventually carried towards the intersection of these lines (although axes are not marked here, this intersection point is $(1, 1)$ by the foregoing scaling and constructions). In this example, the terminal point is both stable and the global attractor. The middle panel of Fig. 5.3 depicts a situation in which sliding motion does not always occur on the singular lines; the portions marked as dashed lines are transversal. Moreover, the intersection is no longer a global attractor, although the point $(1, 1)$ retains part of the plane as its basin of attraction, and trajectories starting in the basin generically have a terminal arc of sliding motion. The right-most panel of Fig. 5.3 depicts a situation in which the point $(1, 1)$ is a global attractor but it is not locally stable, since perturbations to the left give rise to orbits that spiral around and then return on a terminal sliding arc.[†] ❖

[†]In Example 5 we evoked the idea of 'systems ætiology' whereby it is the eventual failure of regulatory compensation that precipitates the acute, fulminant clinical manifestation of a disease. The present discussion of signatures adds to this notion, as follows. First, whether or not the 1-dimensional intersection point of the singular subsets of phase space has the character of a stable global attractor of the flow (as in the leftmost panel of Fig. 5.3 but not in the other two cases shown) depends on the range of values available

5.3 HIERARCHY OF INTERSECTING DISCONTINUITY HYPERSURFACES

The intersection $\cap_{j=1}^{m} \mathcal{L}_j^\circ$ is an affine space in $\mathbb{R}^n$ that can be described as the null space of the $m \times n$ matrix $\boldsymbol{\kappa}$ translated by the vector $\mathbf{1}$. The homeostatic point $\boldsymbol{x} = \mathbf{1}$ is in this intersection, and if we demand that this be its *only* element, the nullity of $\boldsymbol{\kappa}$ must be zero, implying that the rank of $\boldsymbol{\kappa}$ must be equal to n. Since $\text{rank}(\boldsymbol{\kappa}) \leq m$, this condition becomes $n \leq m$. If the system does not satisfy this requirement, it is *homeostatically deficient*. In intuitive terms, there are not enough actuators to manage the requirements of the (fully singular) homeostatic state.

With each $\mathcal{L}_j^\circ$ we associate a unit vector $\boldsymbol{v}_j^\circ$ that is orthogonal to $\mathcal{L}_j^\circ$ and points in the direction where z_j is stopped at its maximum Anschlag value $z_{j,\max}$. These m vectors can be linearly independent only if $n \geq m$; and only if they are linearly independent can their span be subdivided into 2^m regions corresponding to one of the 2^m possible ways in which z can be fully stopped (i.e. every z_j taking either the value $z_{j,\min}$ or $z_{j,\max}$). If the system does not satisfy this requirement, it is *homeostatically redundant*. In intuitive terms, there are too many actuators, leaving the singular values of z underdetermined by the homeostatic requirements encoded in L in dependence of $\boldsymbol{x}$. In the present context of DIs, we suppose that L has no additional dependence on any given $\mathcal{Z}\sim$variable, provided it is bounded away from a lower or upper stop value; in general, such additional dependences may settle the problem of homeostatic redundancy (cf. Fig. 3.2).

Systems satisfying $n = m$ are neither homeostatically deficient ($m < n$) nor homeostatically redundant ($m > n$). In such systems, the vectors $\{\boldsymbol{v}_1^\circ, \ldots, \boldsymbol{v}_n^\circ\}$, provided that they are linearly independent, form an alternative basis for $\boldsymbol{x}$. In particular, assume such linear independence and let $\boldsymbol{x} = \left[\boldsymbol{v}_1^\circ \cdots \boldsymbol{v}_n^\circ\right] \cdot \widetilde{\boldsymbol{x}}$ define the transformed state $\widetilde{\boldsymbol{x}}$, which inhabits a phase space partitioned into hyperoctants that are bounded by the subspaces corresponding to $\{\mathcal{L}_j^\circ, \ldots, \mathcal{L}_n^\circ\}$. Each such hyperoctant corresponds to one way in which z can be fully stopped (Fig. 5.4).

Letting d_0 denote the number of elements of the transformed state vector $\widetilde{\boldsymbol{x}}$ that are zero, we see that $n - d_0$ is the effective dynamical

to the $\mathcal{Z}\sim$variables. Transversals become possible when this range does not permit these variables to take on the singular values that suffice to keep the phase point confined to a singular subset. In themselves, transversals are not necessarily indicative of 'unhealthy' physiology; however, any change from an attractive character (signature $\{+,+\}$) to a transversal (signatures $\{+,-\}$, $\{-,+\}$) or a repelling ($\{-,-\}$) character may be a marker of 'subclinical homeostatic insufficiency.'

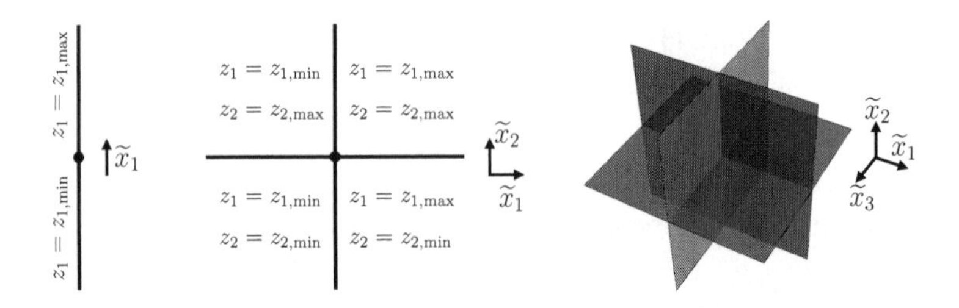

Figure 5.4 **Partitioning of the state space in hyperoctants.** Left: the case $n = m = 1$; The hyperoctants are the positive and negative halves of the $\widetilde{x}_1$-axis, with the Anschlag values of the associated z_1 indicated, and the discontinuity subspace is the point $\widetilde{x}_1 = 0$. Middle: the case $n = m = 2$; The hyperoctants are the quadrants of the $(\widetilde{x}_1, \widetilde{x}_1,)$-plane, with the Anschlag values of (z_1, z_2) indicated, and the discontinuity subspaces are formed by the two axes and their point of intersection. Right: the case $n = m = 3$; The hyperoctants are the octants in the regular sense and the discontinuity subspaces are formed by the indicated planes and their intersections.

dimension of the system (i.e. $n - d_0$ is the number of dynamic degrees of freedom). For $d_0 = n$, the system is at its homeostatic point, $\boldsymbol{z}$ is at the singular value required to keep the system there (under the prevailing value of $\boldsymbol{u}$), and no additional first-order ODEs are required to describe the dynamics. At the other extreme, if $d_0 = 0$, $\boldsymbol{z}$ is fully stopped (in one of the 2^n possible configurations) and the number of first-order ODEs required is n, to describe the movement in the interior of one of the hyperoctants. In general, d_0 of the elements of $\boldsymbol{z}$ have singular (non-stopped) values and $n - d_0$ ODEs are necessary to describe the dynamics of the system. In terms of the original basis, we see that $n - d_0$ out of the n original ODEs in the system $\dot{\boldsymbol{x}} = \boldsymbol{f}$ are retained, while d_0 of the elements of $\boldsymbol{x}$ are fixed algebraically by the fact that $\boldsymbol{x}$ is in the intersection of d_0 of the $\mathcal{L}_i^\circ$ (and the elements are not governed by an ODE).

From the DI point of view, the behaviour of $\boldsymbol{x}$ consists of movement of the phase point described by $\sum_{d_0=0}^{n} \binom{n}{d_0} = 2^n$ distinct systems of equations, each associated with a region of the phase space $\mathcal{X}$. One of these is fully algebraic, with n equations fixing the n $\mathcal{Z}$~variables at singular values and $\boldsymbol{x} = \mathbf{1}$.

The other extreme is the fully dynamic case. Here we have n ODEs of the form $\dot{x}_i = f_i$ and the state $\boldsymbol{x}$ is travelling through the interior of one of the hyperoctants. Each of the $\mathcal{Z}$~variables is stopped at one of its extreme values.

The idea of formulating distinct sets of governing equations (algebraic, differential, or otherwise) on the various regions in a partitioning of the state space has a checkered past, since even experienced mathematical modellers may fall foul of unexpected inconsistencies and side effects when approaching such a task on an *ad hoc* basis. It is therefore with some justification that 'patch-work' models, as well as models with hard non-linear transitions, have long been regarded as examples of bad practice. Matters truly are different in the present case, where the patchwork arises *systematically* out of smooth and globally uniform dynamics on $\mathcal{X} \times \mathcal{Z}$, without any 'stitches.'

5.4 COMPARISON TO THE FILIPPOV AND UTKIN CONSTRUCTIONS

The differential inclusion $\dot{\boldsymbol{x}} \in \boldsymbol{F}(\boldsymbol{x})$ stipulates that the rate of change of $\dot{\boldsymbol{x}}$ be an element of the set-valued map

$$\boldsymbol{F}(\boldsymbol{x}) = \{\boldsymbol{f}(\boldsymbol{x}, z)\}_{z \in \mathcal{Z}} \tag{5.6}$$

(i.e., a *parametrized set-valued map* [6]; we omit $\boldsymbol{u}$ in this section, assuming it fixed). The problems that immediately arise in this setting are that of selecting an element from $\mathcal{Z}$ at each point in time t and of ascertaining the uniqueness and differentiability of the system trajectory thus induced [6]. The first problem has been our main concern, since it is precisely that of choosing controls, whereas we have largely passed by the second problem.

The problem of choosing controls has been couched in terms of finding a function $\overline{\boldsymbol{g}}$ so that $z = \overline{\boldsymbol{g}}(\boldsymbol{x})$ and $\boldsymbol{F}$ becomes $\widetilde{\boldsymbol{f}}$, the latter being single-valued except possibly in special regions where $\overline{\boldsymbol{g}}$ is multi-valued. Such problems can be formulated as piecewise-smooth dynamic systems [20]:

$$\dot{\boldsymbol{x}} = \begin{cases} \boldsymbol{F}_1(\boldsymbol{x}), & H(\boldsymbol{x}) > 0 \\ \boldsymbol{F}_2(\boldsymbol{x}), & H(\boldsymbol{x}) < 0 \end{cases} \tag{5.7}$$

where H is a smooth function and the discontinuity boundary $\boldsymbol{\Sigma}$ is defined to be the locus of $H(\boldsymbol{x}) = 0$. In terms of the physiological Lagrangian, we might consider a linking formula such as $H = \prod_{i=1}^{n} \partial L / \partial x_i$.

Sliding motions on $\boldsymbol{\Sigma}$ are possible within a domain $\mathcal{D} \subseteq \mathcal{X}$ if for all $\boldsymbol{x} \in \boldsymbol{\Sigma} \cap \mathcal{D}$ the difference $\boldsymbol{F}_1 - \boldsymbol{F}_2$ has the same degree of smoothness, which is defined as 1 plus the leading order of the partial derivative of this difference, evaluated on Σ [20]. The central importance of sliding motion in the analysis of biological regulation has been observed by several authors (e.g. refs. 17, 14, and 62), the essential idea being confinement of the organism's state to a subset of $\mathcal{X}$ whose elements are states compatible with life.[‡]

Sliding motion may be attracting, transversal, or repelling, the former being of most interest [20]. The selection problem in this case has been addressed by both Filippov [23] and Utkin [84]. Filippov [23] takes a convex combination of the flow fields; that is, on Σ the dynamic law is $\widetilde{f} = (1-\alpha)\boldsymbol{F}_1 + \alpha\boldsymbol{F}_2$ with $\alpha \in [0,1]$ chosen so as to keep the flow vector tangential to Σ. Utkin [84], on the other hand, gives $\widetilde{f}$ as the average of $\boldsymbol{F}_1$ and $\boldsymbol{F}_2$ plus $(\boldsymbol{F}_2 - \boldsymbol{F}_1)/2$ times a control variable $\beta \in [-1,+1]$. These formulas are obviously equivalent (with $\beta = 2\alpha - 1$), but Filippov's approach recalls the intimate connection of DI theory with convex analysis whereas Utkin's stresses the link to optimal control theory.

In the present formulation, the extreme points of the values allowed for the control correspond to the smooth flows on either side of the boundary. However, since the flow depends on the control in a generally non-linear manner, Eq. (5.6), the singular control value that keeps the flow contained in Σ need not result in a flow that is a convex combination of its extremes — even if many realistic models turn out be close or identical to the Flippov/Utkin selection.

Moreover, we have proposed a dynamic 'smoothification' of the DI, in particular, a generalised version of integrating control that rules the $\mathcal{Z}$~variables. Naturally, the Flippov/Utkin formalism is readily extended in like manner. All that is required is the elevation of α or β to the status of state variable, endowed with a dynamic law such as the following:

$$\dot{\beta} = -\mu H(\boldsymbol{x}) - \frac{\beta}{(1-\beta)(1+\beta)} \tag{5.8}$$

where β is Utkin's control and μ is a positive parameter.

[‡]We think here of functional and structural integrity. The intuitive notion that the life-compatible subset of $\mathcal{X}$ must be in some sense special, surely minute compared to $\mathcal{X}$ itself, resists being made more precise. Stating that it is of measure zero in $\mathcal{X}$ is somewhat trite (and may not even be true, depending on how one specifies the 'parent' space $\mathcal{X}$). We might surmise that reduced entropy is key, given that entropy is likewise a matter of tiny subregions (from the perspective of statistical physics), and that biological organisation necessarily involves entropic reduction (organisation in the sense of cellular architecture, tissue organisation, anatomy, nervous connectome). However, the latter is astronomically tiny compared to the magnitude of ΔS due to processes occurring below the macro-molecular level of organisation [12]. It is tempting to speculate that, for the state space partitioning that we seek here, a quantity that plays an analogous role to entropy is the one defined by Eq. (7.9).

CHAPTER 6

Application to mammalian nutrient budgets

The theory developed in the present monograph aims to understand the effects of conflicts that arise between homeostatic drives when regulatory loops intersect, and to elucidate certain aspects of organismal physiology by viewing them through the lens of such 'interlocking loops.' Several tools have been developed in the foregoing chapters; the present chapter aims to bring these all to bear on a fairly substantial case study. The nutrient and energy budgets in mammals constitute a system of choice: the interlocking loops involve the interactions between a 'glycostat,' a 'myostat,' and a 'lipidostat.' There is a somewhat subtle biochemical rationale for the concurrent existence of three such 'stat' functions. The 'stats' are subject to a hierarchy of stringency of set-point protection (which we may define as the magnitude of $\partial^2 L/\partial x_i^2$ at the physiological optimum). This hierarchy informs how the organism responds to starvation, malnutrition, and overfeeding. The DI formalism plays a key role in visualising the phase flow of the physiological component.

The mathematical theory of the physiological component will be based on basic physico-chemical principles, as we advocated in Chapter 1. In particular, we will base our approach on the nutrient and energy budget theories proposed by Pütter [71], Von Bertalanffy [85] and, more recently, Kooijman [47]. These models all depart from the assumption that, for a given organism, a length measure x_L can be defined in such a way that assimilation of a nutrient is proportional to x_L^2 and fluxes associated with space-filling processes, such as maintenance expenses, are proportional to x_L^3. This fundamental assumption is known as the *Flächenregel* [68].

In the Pütter-Bertalanffy-Kooijman framework, dynamics of organismal growth are derived by combining the Flächenregel with chemical conservation principles. This results in an ODE for x_L as well as ODEs describing the 'densities' of nutrient reserves. In the following section, we review the budget framework in general terms.

6.1 THE DYNAMIC NUTRIENT BALANCE

A dynamic balance can be made up for any particle species of interest, denoted '$\star$' in general. Examples include the biogenic elements (C, H, O, N, S, P, ...), functional groups or the oxidation states of a given element. These groups could be basic ones such as $-NH_2$, -COOH, or more complex groups such as vitamins, co-factors. etc. In the case where oxidation states are distinguished, one particular chemical element is split into several distinct species of interest. As we shall presently see, mammalian metabolism necessitates the division of carbon into a low and high oxidation state.

Let $N_\star$ denote the number of particles in the animal and let $U_\star$ be the maximum assimilation flux per unit bio-area x_L^2. In general $U_\star$ must be treated as a function of time t, since food availability depends on environmental factors that are themselves time-varying. Finally, let $z_F \in [0, 1]$ be a feeding multiplier, corresponding to the animal's feeding drive (*appetence*), and let $\phi_\star > 0$ be the expenditure of $\star$ per unit bio-volume x_L^3. This expenditure may involve a biochemical transformation, a flux of matter from the organism to the environment, or both. The balance equation is a straightforward matter of incomings less outgoings:

$$\dot{N}_\star = z_F U_\star x_L^2 - \phi_\star x_L^3 \tag{6.1}$$

where the dot denotes differentiation with respect to time. This equation may also be written in terms of the density $n_\star = N_\star x_L^{-3}$:

$$\dot{n}_\star = \frac{z_F}{x_L} U_\star - \phi_\star - 3\frac{\dot{x}_L}{x_L} n_\star \,. \tag{6.2}$$

In the special case where z_F, $U_\star$, $\phi_\star$, and $n_\star$ are all constant in time (i.e., constant environmental conditions, density homeostasis), Eq. (6.2) yields the classic Pütter-Von Bertalanffy equation:

$$\dot{x}_L = \frac{z_F U_\star}{3n_\star} - \frac{\phi_\star}{3n_\star} x_L \tag{6.3}$$

which, provided that $n_\star > 0$, has solution

$$x_L(t) = \overline{x}_L + (x_L(t_0) - \overline{x}_L)\exp\{\phi_\star(t_0 - t)/(3n_\star)\} \tag{6.4}$$

where $\lim_{t\to\infty} x_L(t) = z_F U_\star/\phi_\star = \overline{x}_L$. Whereas x_L relaxes exponentially, several quantities of interest exhibit a sigmoid dependence on time. This is explained by their scaling as powers of x_L with exponents greater than 1. Body weight, for instance, is proportional to x_L^3.

Reserve densities Eco-physiological considerations lead us to treat $N_\star$ as the sum of two terms: one term associated with the subsistence minimum, defined as the minimum amount compatible with life, and a remainder, which can be regarded as a 'surplus' although not necessarily in the sense of an inert store; it may be tied up in functional machinery, cf. ref. 76. Let us regard the term associated with the subsistence minimum as a function of x_L, thus: $N_\star^\circ(x_L)$. The meaning of 'subsistence' is precisely that remaining alive is compatible only with the condition $N_\star \geq N_\star^\circ(x_L)$, death becoming unavoidable as soon as $N_\star < N_\star^\circ(x_L)$.

The particle species is *non-essential* iff $N_\star^\circ(x_L) = 0$, whereas *essential* particle species are characterised by $N_\star^\circ(x_L) > 0$. We shall focus on the case where the subsistence minimum $N_\star^\circ(x_L)$ is proportional to the biologically active volume of the organism, which scales as x_L^3. We thus write

$$N_\star^\circ(x_L) = n_\star^\circ x_L^3 \tag{6.5}$$

where $n_\star^\circ$ is a positive constant, called the *subsistence coefficient*. The difference $N_\star - n_\star^\circ x_L^3$ represents a 'surplus' or a 'reserve' (with the caveat mentioned above).

Let there be n stoichiometrically distinct storage compounds and choose n particle species of interest. Let $\boldsymbol{C}$ be the $n \times n$ matrix whose columns represent the compositions of the storage compounds in terms of the particle species of interest, where no column of $\boldsymbol{C}$ has all zero elements. Then we can define storage densities as follows:

$$\boldsymbol{r}_\star = \boldsymbol{C}^{-1} \cdot (\boldsymbol{n}_\star - \boldsymbol{n}_\star^\circ) \tag{6.6}$$

where $\boldsymbol{n}_\star$ collects the densities $n_\star$ (over the n particle species) and $\boldsymbol{n}_\star^\circ$ similarly collects the subsistence coefficients. The rate of change of this vector can be derived on the basis of Eq. (6.1). However, it will prove to be more convenient to work with the dynamics of a state vector $\boldsymbol{x}_\star$ which we define as follows:

$$\boldsymbol{x}_\star = \boldsymbol{C}^{-1} \cdot \boldsymbol{n}_\star\,. \tag{6.7}$$

Since $\boldsymbol{r}_\star = \boldsymbol{x}_\star - \boldsymbol{C}^{-1} \cdot \boldsymbol{n}_\star^\circ$ and $\boldsymbol{C}^{-1} \cdot \boldsymbol{n}_\star^\circ$ is constant, the rates of change are identical, i.e. $\dot{\boldsymbol{r}}_\star \equiv \dot{\boldsymbol{x}}_\star$. To obtain the latter, let $\boldsymbol{u} = \boldsymbol{C}^{-1} \cdot \boldsymbol{U}$ where $\boldsymbol{U}$ collects the values of $U_\star$ over the n particle species. Similarly, let $\boldsymbol{\phi}$ collect the expenditure fluxes $\phi_\star$ over the n species and let $z_L = x_L^{-1} \dot{x}_L$ denote the relative growth rate. Then Eq. (6.2) becomes:

$$\dot{\boldsymbol{x}}_\star = z_F x_L^{-1} \boldsymbol{u} - \boldsymbol{C}^{-1} \cdot \boldsymbol{\phi} - 3 z_L \boldsymbol{x}_\star \,. \tag{6.8}$$

This equation combines the *Flächenregel* with the stoichiometric mass balance requirement (which is satisfied by every chemical and biochemical reaction). The subsistence requirement can be written as

$$\boldsymbol{x}_\star \geq \boldsymbol{C}^{-1} \cdot \boldsymbol{n}_\star^\circ \tag{6.9}$$

where the symbol $\geq$ should interpreted element-wise.

Assimilation Different species of particles are ingested in fixed ratios depending on the compositions of the foodstuffs, which can be represented as the columns of a matrix $\boldsymbol{B}$. If $\widetilde{\boldsymbol{U}}$ collects the maximum specific ingestion rates of the foodstuffs, then we have $\boldsymbol{U} = \boldsymbol{B} \cdot \widetilde{\boldsymbol{U}}$. This conversion is relevant in the context of food chain (web) theory: the elements of $\widetilde{\boldsymbol{U}}$ relate to the various food or prey items in the animal's surroundings (e.g. via Holling-type response curves, or generalizations thereof [10]). Food items can be stoichiometrically similar to storage polymers. Carnivores, for instance, consume animals of a similar biochemical nature, and the prey's composition is a linear superposition of stoichiometries similar to the storage polymers of the predator. Herbivores have dietary items similar to one of the storage polymers, e.g., starch corresponding to glycogen. In such cases, we are able to describe the intake flux $\widetilde{\boldsymbol{U}}$ in such a way that $\boldsymbol{B} \approx \boldsymbol{C}$ holds.

6.2 MINIMAL MODEL FOR THE MAMMALIAN SYSTEM

Nutrient disposition in mammals is severely constrained by the fact that mammals cannot synthesise glucose from fatty acids [77]. This implies that the body's carbon content has to be divided into two classes: *glucogenic* carbon which is interconvertible into any of the *gluconeogenic precursors* (i.e., glucose and other glycolytic metabolites, as well as carbohydrates such as glycerol and glycerol-3-phosphate), and *ketogenic* carbon which is interconvertible into acetyl-CoA and its 'polymeric storage' form, the fatty acids. Physically, carbon atoms are in fact capable of moving from the ketogenic pool to the glucogenic pool, but for every carbon atom that is transferred, one carbon atom is obligatorily oxidized into CO_2, and thus there can be no net transfer. (It has been claimed that mammals express pathways which can metabolise certain fatty acids to gluconeogenetic precursors, but this is thought to be of virtually no physiological importance [21, 77].)

The ketogenic pool cannot furnish carbons for the glucogenic pool as far as nutrient budgets are concerned. Conversely, however, it is possible for the glucogenic carbon pool to feed the ketogenic pool: the biochemical reaction that is catalysed by the enzyme pyruvate dehydrogenase transforms glucogenic carbon into ketogenic carbon [58]. In sum: net gluco- to keto- conversion is possible, but not *vice versa.*†

---

†It might be thought convenient if the organic compounds of mammalian metabolism could all be assigned to either the glucogenic or the ketogenic side, but, as a result of the keto/gluco conversions that are allowed, plus the fact that the Krebs cycle is a central hub for amino acid and nucleoside metabolism [58], this is not the case.

Mammals contain carbon in both organic and inorganic form; the latter is not included in the particles of interest (its mass is simply assumed to be proportional to x_L^3; i.e. it is accounted for in the same way as other elements not included among the particles of interest, such as H and O). In general, the carbon atoms in each metabolite can be accorded to either one of the pools by inspection of the metabolic pathways [58]. The distinction of glucogenic and ketogenic carbon is in fact a coarse graining of a far greater variety of carbon 'species' defined by the atoms' chemical environment (e.g. carbon oxidation states vary from $-4e^-$ to $+4e^-$), which entails slight inaccuracies.

Acetate, as well as the ketone bodies acetoacetate and D-3-hydroxybutyrate, yield acetyl-CoA and are thus ketogenic. Triacylglycerols are almost entirely ketogenic; e.g. tripalmitin is $48/51$ ketogenic: its fatty acids yield acetyl-CoA whereas its glycerol is glucogenic. Phospholipids are slightly more glucogenic; e.g. stearyl-*cis*-oleiyl phosphatidylcholine is $36/44$ ketogenic.

The glycolytic metabolites as well as alanine, glycine, serine, and cysteine are fully glucogenic. The Krebs cycle intermediates succinyl-CoA, succinate, fumarate, malate, and oxaloacetate are $1/4$ ketogenic, since 1 carbon atom is lost as carbon dioxide in gluconeogenesis — when glucose is generated from pyruvate via dicarboxylates, an inorganic carbon is condensed and later evolved, allowing pyruvate to make a net contribution of 3 carbon atoms to each glucose molecule, which renders pyruvate fully glucogenic. Similarly, citrate and isocitrate are $3/6 = 1/2$ ketogenic, whereas oxoglutarate is $2/5$ ketogenic. Krebs cycle-derived amino acids inherit their ketogenicity from their precursors: thus, glutamate, glutamine, proline, and ornithine are $2/5$ ketogenic whereas aspartate and asparagine are $1/4$ ketogenic. Citrulline and arginine are both $2/6 = 1/3$ ketogenic and $3/6 = 1/2$ glucogenic; two ketogenic carbons are inherited from ornithine and one carbon comes from carbamoyl phosphate, which we treat as inorganic carbon 'in transit.'

For essential amino acids, the catabolic pathways determine keto/glucogenicity. Threonine is $1/4$ ketogenic, evolving a carbon dioxide molecule as it is metabolised to pyruvate. Histidine is $2/6 = 1/3$ ketogenic: its catabolism yields glutamate ($2/5$ ketogenic) and a single carbon transferred via N^5,N^{10}-methenyl tetrahydrofolate (part of the 1-carbon pool) to glycine, which yields serine, which can be deaminated to pyruvate. Tyrosine and phenylalanine are $6/9 = 2/3$ ketogenic since they yield fumarate plus the ketone body acetoacetate, one carbon being lost as carbon dioxide. Lysine is fully ketogenic as it yields acetyl-CoA, as well as evolving two carbon dioxide molecules. Tryptophan is 7/11 ketogenic, yielding glucogenic alanine (3 C) and formate (which joins the 1-carbon pool), plus ketogenic acetyl-CoA (2×2 C) and 3 carbon dioxide molecules. Methionine is 1/5 ketogenic as it is metabolised to succinyl-CoA, further yielding a methyl group that joins the 1-carbon pool; the pathway also evolves a carbon dioxide molecule and condenses one.

The branched-chain amino acid degradation pathways both evolve and condense carbon dioxide. Valine is 2/5 ketogenic; its catabolism involves the evolution of 2 carbon dioxide molecules, while it condenses one, yielding succinyl-CoA. The catabolism of isoleucine evolves and condenses one carbon dioxide, yielding succinyl-CoA and acetyl-CoA, which makes isoleucine 3/6=1/2 ketogenic. The catabolism of leucine also evolves and condenses one carbon dioxide, but yields acetyl-CoA and acetoacetate which makes leucine fully ketogenic.

The energetic requirements are represented in the minimal model as loss term in the dynamics of ketogenic carbon. This correlation of energy generation with loss from the ketogenic carbon is generally warranted since complete oxidation of organic carbon in mammalian metabolism requires the conversion of glucogenic to ketogenic carbon by pyruvate dehydrogenase. However, fluctuations in the coupling between energy generation and ketogenic loss occur on a short time scale, as a result of anaerobic energy generation, which produces glucogenic lactate that is subsequently either further catabolised or used as a substrate for gluconeogenesis (fueled by oxidation of ketogenic carbon). These 'oxygen debt' fluctuations average out over time.

Furthermore, the stoichiometry between ATP or reducing equivalents and carbon is somewhat variable. Fatty acid oxidation yields slightly over 7 ATP per ketogenic carbon (the exact number depending on the length of the acyl tails) whereas glucose yields only 5 1/3 ATP per glucogenic carbon. This deficit of 5/3 ATP per carbon atom is due to differences in substrate-level phosphorylation; it can be defrayed by a loss term $-\sigma z_{GK}$. This calculation assumes that all conversion of glucogenic to ketogenic carbon is dissimilatory; in reality, there is an assimilatory contribution whenever ketogenic reserves are growing. However, this reserve synthesis incurs an energetic cost of the same approximate magnitude per carbon converted, and to a good approximation the same σ can be used to represent dissimilatory and assimilatory gluco/keto conversions. The parameter $\kappa = -\sigma + 2/3$ combines the stoichiometric conversion constant 2/3 (as pyruvate loses a carbon when transformed into acetyl-CoA) with the substrate-level correction factor σ; it would seem reasonable to assume $\kappa \approx 1/2$.

Glycogen or acyl-derived carbon serves as the main fuel under most conditions. However, muscle protein can also serve as an energy source. In general, both the glucogenic and the ketogenic carbons in amino acids have a lower ATP yield.

Certain organic compounds cannot be catabolised completely to the intermediates of core metabolism and are instead excreted as complex derivates (urate/allantoin, bile pigments, cholates). This class includes purines, porphyrins, as well as various messenger molecules (e.g. catecholamines, steroid hormones) and co-factors. It seems most expedient to treat the carbon atoms in these compounds as ketogenic (their ATP-yielding capacity is small or zero, somewhat reducing the ATP-equivalency of the average ketogenic carbon). This allows us to consider their turn-over as included as a (very minor) component of the maintenance requirements. The biosynthesis of such compounds often involves glucogenic precursors; for instance, the purines guanine and adenine both contain 5 carbon atoms, of which 4 derive from glucogenic donors (glycine and the 1-carbon pool), the remaining one from the inorganic carbon pool. There is thus a biosynthetic flux of glucogenic carbon into the non-degradable pool, which contributes (to a very small measure) to z_{GK}.

The pyrimidines can be fully catabolised, although normal nucleoside turn-over salvages the bases rather than fully catabolising them. Thymine presents no difficulties, as it is found to be 2/5 ketogenic both in terms of its precursors and its catabolic products, which are

Important limiting factors in mammals are the availabilities of energy-furnishing compounds and nitrogen-rich foodstuffs. A minimal nutrient budget model for mammals must therefore account for three particle species: nitrogen (N), glucogenic carbon (C_G), and ketogenic carbon (C_K); hydrogen and oxygen are not explicitly accounted for in the minimal model. Bearing this in mind, we specify the quantities introduced in Section 6.1, beginning with the stoichiometric reserve matrix:

$$\boldsymbol{C} = \begin{bmatrix} 1 & 0 & 0 \\ C_{GN} & 1 & C_{GK} \\ C_{KN} & 0 & 1 \end{bmatrix} \tag{6.10}$$

where the first column represents the composition of the nitrogen reserves, mainly muscle protein. The elements of this column are relative to the amount of carbon stored, so C_{GN} is the number of glucogenic carbon atoms per carbon atom stored, and similarly for C_{KN} (this scaling accounts for the fact that the diagonal elements are all unity). The second column is $[0, 1, 0]^T$, since the primary storage polymer for glucogenic carbon, glycogen, contains neither nitrogen nor ketogenic carbon. The third column of $\boldsymbol{C}$ reflects the composition of lipid storage (predominantly triacylglycerol) which contains no nitrogen and a minor contribution C_{GK} from glucogenic carbon.

Expenditures The flux $\boldsymbol{\phi}$, which collects physiological expenditures, is specified as follows:

$$\boldsymbol{\phi} = \begin{bmatrix} 1 & 0 & 0 & 0 \\ 0 & 1 & 0 & 1 \\ 0 & 0 & 1 & -\kappa \end{bmatrix} \cdot \begin{bmatrix} z_{NX} \\ z_{GX} \\ z_{KX} \\ z_{GK} \end{bmatrix}, \tag{6.11}$$

where z_{NX}, z_{GX}, and z_{KX} denote the rates at which N, C_G, and C_K are excreted, respectively, and z_{GK} represents the rate of conversion of C_G to C_K. Hence the first three columns of the matrix make up an identity matrix. The last column reflects the fact that κ C_K atoms are gained for every C_G atom lost ($\kappa < 1$). Mammals excrete nitrogen in the form of ammonia and urea, which involve other elements besides nitrogen [26]; however, hydrogen and oxygen (or any of the other biogenic elements) are not included in the minimal model budget, as they can be regarded as non-limiting under virtually all circumstances, and the carbon in urea is of

succinyl-CoA and carbon dioxide. On the other hand, cytosine and uracil yield malonyl-CoA and carbon dioxide when fully catabolised. The net effect of one cycle of synthesising and catabolising either cytosine or uracil is the conversion of 3 glucogenic carbons into 1 ketogenic carbon and 2 carbon dioxide molecules.

inorganic origin; that is, as ketogenic carbon is respired, it leaves the budget in the form of carbon dioxide and a small portion of this carbon dioxide 'on its way out' is combined with ammonia to be excreted as urea [58]. Creatinine is produced in proportion to muscle mass ($\sim r_{\mathrm{N}}$) and accounts for obligatory losses of N, $\mathrm{C_G}$, and $\mathrm{C_K}$; that is, there are minimum positive lower bounds (lower stops) to, respectively, z_{NX}, z_{GX}, and z_{KX} [96]. Use of glucogenic carbon compounds (principally glucose) to regenerate Gibbs enthalpy carriers, such as ATP, is part of the flux z_{GK}; as the glucogenic carbon atoms pass into the Krebs cycle via pyruvate dehydrogenase, they become part of the ketogenic side of metabolism [58]; once ketogenic, the carbon atoms can pass back and forth between the Krebs cycle and the fatty acid pool, but not return to the glucogenic side of metabolism. Several tissues obligatorily oxidise carbohydrates and thus rely on gluco/keto conversion to regenerate Gibbs enthalpy [26]. This implies a non-zero, positive lower bound on z_{GK}; however, even such 'glucose-dependent' tissues may switch to alternative fuels of (partly) ketogenic nature under prolonged starvation [26].

The actuator z_{KX} is composed of various terms, which differ in their scaling properties. Basic maintenance metabolism, covering turn-over of the proteome, nucleic acid repair, transport, and maintenance of ionic gradients across membranes, will scale as x_L^3 and thus be represented by a constant contribution to z_{KX}. From a physical point of view, the energetic endpoint is ultimately one of heat dissipation. Homoiotherms lose heat in proportion to x_L^2 (body surface area), and if these losses are not compensated for by the dissipatory losses corresponding to maintenance term, there is additional term in z_{KX} of the form $\max\{0, \alpha(T_c - T_u)x_L^{-1} - \alpha_0\}$ where α_0 and α are positive constants, T_c is the core body temperature, and T_u is the ambient temperature. On the other hand, if maintenance-related heat production exceeds the heat production needed to compensate for losses to the environment ($\alpha_0 x_L^3 > \alpha(T_c - T_u)x_L^2$), the animal has to expend energy to cool down, resulting in a contribution $\max\{0, \alpha_0' - \alpha'(T_c - T_u)x_L^{-1}\}$ where α_0' and α' are positive constants. The pernicious subtlety here is that $\alpha \neq \alpha'$ in general [14]; otherwise the offset could be incorporated into the constant term and a term proportional to x_L^{-1} would cover heat balance in either case. Additional terms in z_{KX} arise in proportion to (sustained, strenuous) muscular activity and, finally, Gibbs-enthalpic requirements related to growth (giving rise to a term $\propto z_L$ which is not to be confused with the final term in Eq. (6.8), as this term is merely the result of the quotient rule of differentiation).

We can rewrite Eq. (6.8) as

$$\dot{\boldsymbol{x}}_\star = z_F {x_L}^{-1}\boldsymbol{u} - \boldsymbol{K}\cdot\boldsymbol{z} - 3z_L\boldsymbol{x}_\star \tag{6.12}$$

where we have introduced the following notation:

$$\boldsymbol{x}_{\star} = \begin{bmatrix} x_{\mathrm{N}} \\ x_{\mathrm{G}} \\ x_{\mathrm{K}} \end{bmatrix}, \quad \boldsymbol{u} = \begin{bmatrix} u_{\mathrm{N}} \\ u_{\mathrm{G}} \\ u_{\mathrm{K}} \end{bmatrix}, \quad \boldsymbol{z} = \begin{bmatrix} z_{\mathrm{NX}} \\ z_{\mathrm{GX}} \\ z_{\mathrm{KX}} \\ z_{\mathrm{GK}} \end{bmatrix}, \tag{6.13}$$

and

$$\boldsymbol{K} = \boldsymbol{C}^{-1} \cdot \begin{bmatrix} 1 & 0 & 0 & 0 \\ 0 & 1 & 0 & 1 \\ 0 & 0 & 1 & -\kappa \end{bmatrix} = \begin{bmatrix} 1 & 0 & 0 & 0 \\ -C_{\mathrm{GN}} + C_{\mathrm{GK}} C_{\mathrm{KN}} & 1 & -C_{\mathrm{GK}} & 1 + \kappa C_{\mathrm{GK}} \\ -C_{\mathrm{KN}} & 0 & 1 & -\kappa \end{bmatrix}. \tag{6.14}$$

The state $\boldsymbol{x}_{\star}$ accounts for the stoichiometry of the system, i.e. the dynamic budget of nitrogen and the two species of carbon. Additional elements can be added to achieve a more complete stoichiometric account. Moreover, further non-stoichiometric $\mathcal{X}$~variables can be added; these do not account for additional N, C_G, or C_K atoms, but describe other aspects of biochemistry and physiology. One such variable is x_L, which has dynamics $\dot{x}_L = z_L x_L$. We resist the temptation to add more $\mathcal{X}$~variables, as our present objective is a minimal model, sufficient to demonstrate the main ideas.

The coupling graph for this system is bipartite and is depicted in Fig. 6.1. The present case is particularly simple, since the couplings are exclusively of the 'κ' type (see Section 4.1) and as a result, the zeroth-order approximation expressed by Eq. (4.4) is exact.

6.3 QUALITATIVE ANALYSIS OF THE MINIMAL MODEL

The ODEs, Eq. (6.12) plus $\dot{x}_L = z_L x_L$, represent the dynamics $\boldsymbol{f}$ of $\boldsymbol{x}$. The next step is to express homeostatic drives in terms of the physiological Lagrangian L that governs the dynamics of z as detailed in Chapter 3. To this end, it suffices to make three fairly innocuous assumptions, the first two of which are primarily made for the sake of simplicity and could readily be relaxed, whereas the third is essential for much that follows, expressing the physiological hypothesis that the set-point of the protein reserves (muscle mass) is more stringently protected than that of the glycogen stores (C_G), which in turn is more stringently regulated than the adipose stores (which are mostly C_K).

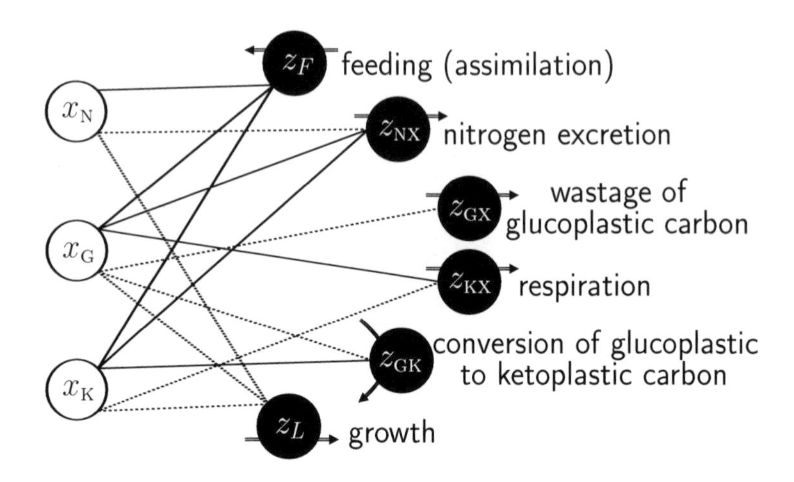

Figure 6.1 **Coupling graph for the mammalian energetics system.** Solid edges represent positive couplings; dashed edges represent negative couplings. Double-shafted arrows indicate the directions of the transformations relative to the x-type nodes.

Without further ado, the assumptions are as follows: (i) the physiological Lagrangian L is *additively separable*, that is, composed of terms each of which corresponds to one of the $\mathcal{X}\sim$ or $\mathcal{Z}\sim$variables, and depends only on that variable; (ii) these terms are, moreover, *convex* functions that *diverge* at the minimum and maximum allowed values of the $\mathcal{X}\sim$ or $\mathcal{Z}\sim$variables; (iii) their minima, located at $x_N = \widehat{x_N}$, $x_G = \widehat{x_G}$, and $x_K = \widehat{x_K}$, satisfy the following condition:

$$\left.\frac{\partial^2 L}{\partial x_N^2}\right|_{x_N=\widehat{x_N}} \gg \left.\frac{\partial^2 L}{\partial x_G^2}\right|_{x_G=\widehat{x_G}} \gg \left.\frac{\partial^2 L}{\partial x_K^2}\right|_{x_K=\widehat{x_K}} . \tag{6.15}$$

The minimum allowed values for $\boldsymbol{x}_\star$ are given by $\boldsymbol{C}^{-1} \cdot \boldsymbol{n}_\star^\circ$ and the maximally allowed values are related to a maximum capacity for the storage of the corresponding biopolymers.

Discontinuity surfaces To understand the qualitative dynamics, let us first consider the discontinuity surfaces in the (x_N, x_G, x_K) phase space. Three of these are depicted in Fig. 6.2: one corresponding to z_{NX} (nitrogen wasting), one corresponding to z_L (growth), and one corresponding to z_{GK}

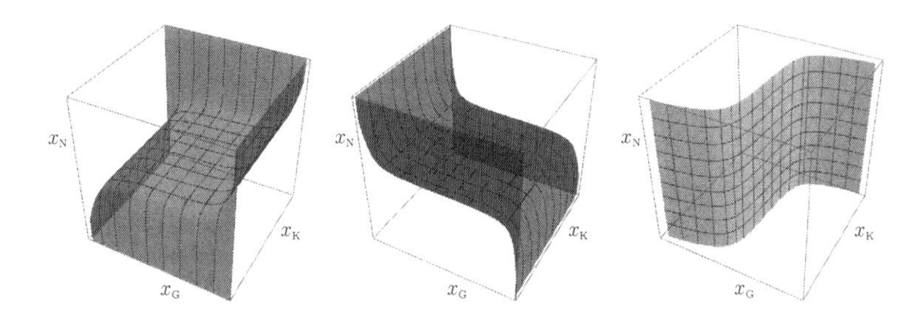

Figure 6.2 **Discontinuity surfaces (singular surfaces) in the mammalian macrochemical body composition phase space.** Left: nitrogen excretion, wasting (z_{NX}); middle: growth (z_L); right: conversion of glucogenic to ketogenic carbon (z_{GK}).

($C_G \rightarrow C_K$ conversion). For the sake of simplicity we temporarily assume that $\mathcal{Z}$~variables other than z_{NX}, z_L, and z_{GK} are held at constant values.

Intersections of the three discontinuity surfaces are depicted in Fig. 6.3. It is an instructive exercise to study the surfaces as shown in Fig. 6.2 and attempt to superimpose them mentally, before turning to Fig. 6.3. It can be seen that the intersections form several discontinuity lines where two surfaces intersect, as well as a unique discontinuity point where all three intersect. Let us call this the *homeostasis point*.

The diagram in Fig. 6.3 is topologically equivalent to the right-most one in Fig. 5.4, but the surfaces are angled in accordance with the hierarchy indicated in Eq. (6.15). The warped appearance near the boundaries results from the divergence of L in these regions.

Each of these surfaces represents a locus of sliding dynamics, with the corresponding $\mathcal{Z}$~variable in the interior of its range of allowed values. Furthermore, each surface partitions the phase space into two regions: one in which this $\mathcal{Z}$~variable is 'stopped' at its minimum and another in which it is 'stopped' at its maximum. Thus, a discontinuity surface also marks the region where the $\mathcal{Z}$~variable assumes its singular values. For this reason we may also refer to it as a *singular surface*.

In the limiting case suggested by the '$\gg$'-signs in Eq. (6.15) the singular surfaces for z_{NX} and z_L tend to the same horizontal plane, the locus of $x_N = \widehat{x}_N$, over almost the entire range, and the singular surface of z_{GK} would become parallel to the x_K-axis over almost the entire range. However, to exhibit the geometry of the dynamics more clearly, we shall employ the view given here — in other words, the operation indicated by the '$\gg$'-signs has not been taken to its extreme. The optimum point

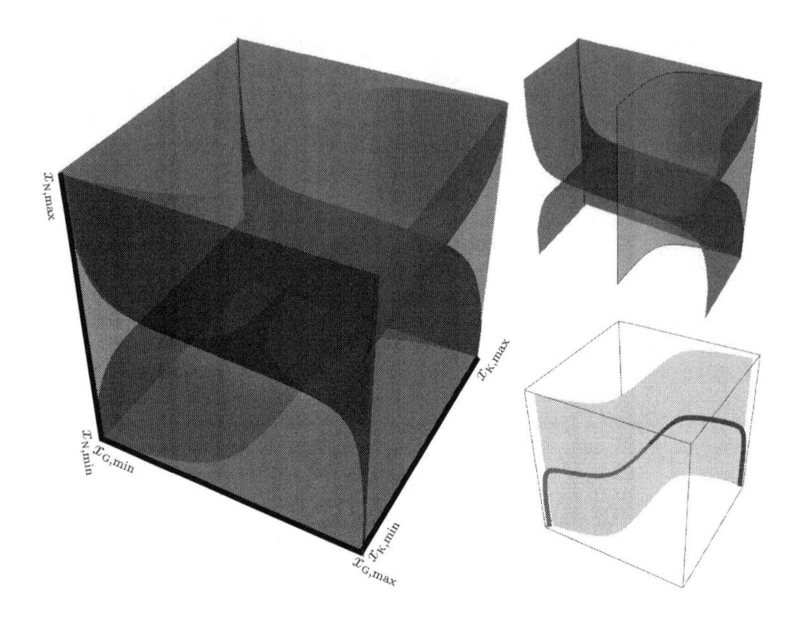

Figure 6.3 **Discontinuity surfaces (singular surfaces): intersections forming discontinuity lines and a discontinuity point.** Left: intersecting system of the singular surfaces for z_{NX}, z_L, and z_{GK}. Right: cut-away versions of the main diagram. Top right: cut through the homeostasis point, i.e., where the three surfaces intersect. Bottom right: the z_{GK} surface, with the intersections with the other two indicated.

$(\widehat{x_G}, \widehat{x_K}, \widehat{x_N})$ has been placed at the centre, again for ease of understanding; in reality we would expect it to be located rather closer to minimal ketogenic reserve density.

Phase flow The qualitative behaviour of the minimal model can be gauged by studying the characteristics of the phase flow. Let us first consider the surfaces as depicted in Fig. 6.4, left panel, which covers the region of phase space where $x_K < \widehat{x_K}$. The singular surface corresponding to z_L here lies above that for z_{NX} (i.e. the left-most surface depicted in Fig. 6.2). Above the singular surface for z_{NX}, the flux z_{NX} assumes its maximum value and if we assume that this value exceeds the maximum possible intake (which is realistic), the phase flow will be downward,

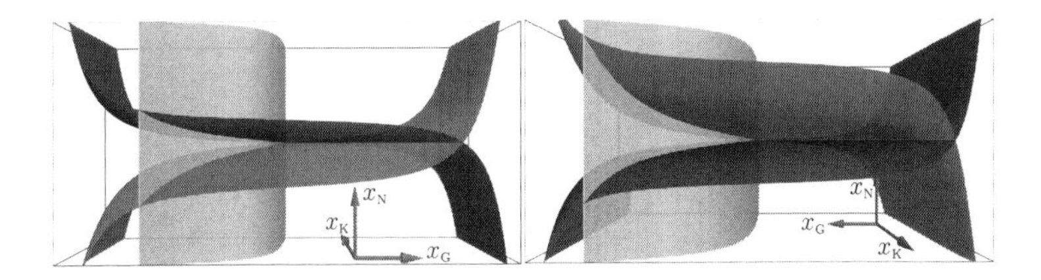

Figure 6.4 **Singular surfaces in the mammalian macrochemical body composition phase space.** Left: detailed view of the region where $x_K < \widehat{x_K}$. Right: detailed view of the region where $x_K > \widehat{x_K}$. The optimum point is at the far end (away from the viewer) in both diagrams.

that is, have a negative component in the direction of the x_N-axis. This implies that the leaf of the singular surface for z_L lying above the singular surface for z_{NX} is a transversal manifold.

Below this leaf, z_L assumes its minimum value (which is 0, i.e. no-growth; we use 'below' in the sense of Fig. 6.4, that is, along the x_N-axis). Furthermore, below the singular surface for z_{NX}, z_{NX} assumes its minimum value. The latter is positive, as normal mammalian physiology is associated with an obligatory loss of nitrogen. However, this obligatory loss is quite small, so that even a meagre diet may generally be expected to supply sufficient nitrogen to cover these outgoings, with at least a little to spare. The phase flow will therefore be upward under normal conditions, back toward the singular surface for z_{NX}.

In sum: *the singular surface for* z_{NX} *attracts the phase flow*, unless the system is experiencing starvation whose severity is such that even minimal obligatory N-losses cannot be replenished; the phase point will detach from the z_{NX} singular surface and roam below it. Barring such conditions, as well as transient phenomena (e.g., recovery from starvation), the dynamics will be confined to the singular surface for z_{NX}, or a boundary region of modest width around it, if the time scale separation (required for the transition to a DI) is incomplete, i.e., the corresponding $\mathcal{L}^{[-\epsilon]}$ and $\mathcal{L}^{[+\epsilon]}$ have not yet merged (cf. Fig. 5.1).

The singular manifold of z_{GK} is vertically oriented in both Fig. 6.2 (right-most panel) and 6.4. To the right of this surface, z_{GK} is set to its maximum value, which is liable to induce a large negative component of the phase flow along the x_G-axis (to the left in the figure).

On the other side of the singular manifold of z_{GK}, the actuator variable z_{GK} assumes its minimum value. Accordingly, the energetic needs are met predominantly from the ketogenic carbon pool. As long as these needs outweigh the intake of C_K, there will be a negative component to the phase flow in the direction of x_K, i.e. towards the viewer in the orientation of Fig. 6.4, left panel. In sum: there are restoring effects on the phase flow on either side of the singular manifold of z_{GK} that will carry the phase point towards this surface.

We conclude that *the singular manifold of* z_{GK} *attracts the global phase flow*. Since we arrived at a similar conclusion as regards the singular surface for z_{NX}, we may infer in addition that any sliding movement along either one of them will tend toward the *intersection* of the two singular surfaces. This intersection is a one-dimensional manifold, as indicated in Fig. 6.3, bottom right.

We have thus far discussed the case $x_K < \widehat{x}_K$ in detail. The case $x_K > \widehat{x}_K$ is similar (Fig. 6.4, right panel). The difference is that here the transversal surface is the singular surface for z_{NX}, which here lies above the singular surface corresponding to z_L. The transversal property obtains if z_L can be set to a maximum value that exceeds the ability for even the richest diet to sustain (this assumption is not met for mammals that lose the ability for somatic growth during the transition from juvenile to adult; below we return to this caveat). Again, we find that the singular manifolds of z_L and z_{GK} attract globally, so that their one-dimensional intersection becomes the arena of the dynamics, barring severe malnutrition or sustained overfeeding events and the recovery transients that may follow such events.

If we concede that the z_{GK}-singular surface acts as a strong global attractor under most feeding regimes (including overfeeding and severe starvation), we are led to consider a simplified phase diagram as depicted in Fig. 6.5. The 1-dimensional manifold defined by the intersection of

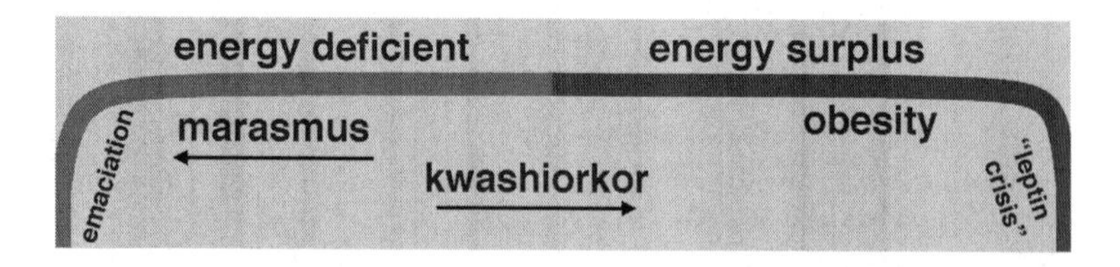

Figure 6.5 **'Regular' dynamics on or in close vicinity to lower-dimensional manifolds.** This diagram should be thought of as an 'unrolled' version of the z_{GK}-singular surface. The optimum point is where the singular surfaces of z_{NX} and z_L meet.

the three singular surfaces (cf. Fig. 6.2 and 6.3, bottom right) acts as a strong attractor under most feeding regimes, moving to the right when the animal is in energy surplus and to the left when the animal is in energy deficiency. To the far left, as the glycogen/lipid-based energy supplies are depleted, protein wasting occurs. The left-ward starvation trajectory has been marked *marasmus* in Fig. 6.5 (μαρασμός, withering), with the final crisis marked *emaciation* — these labels emphasise how the phase flow of the minimal model relates to the standard nomenclature of the corresponding physiological phenomena.

The emaciation section is characterised by a sharp dip in x_N, that is, failure of the myostat. Wastage of muscle mass (which represents the body's reserves of protein) is known as *hypercatabolism* or also *sarcopenia* (σαρκ-, flesh, -πενία, poverty, lack), Accordingly, we may refer to this section of the one-dimensional manifold as a *sarcopenic crisis*. The graph indicates that crisis occurs when x_K nears its minimum allowed value (i.e., near-total depletion of the fat reserves). This 'running together' (=σύν:δρόμος) of fat reserve depletion and muscle mass wastage is known as *metabolic syndrome*.

There will often be a re-feeding phase following severe, prolonged starvation. The phase flow may temporarily become two-dimensional, in the sense that the phase point detaches itself from the one-dimensional manifold and enters other regions on the singular surface of z_{GK}. Physiologically, this reflects a failure of x_N to 'keep up,' which may happen when the refeeding diet is too rich in C relative to N. When infants are weaned and subsequently fed a diet rich in carbohydrates (or fat) but poor in protein, the excessive C/N ratio leads to a malnutrition syndrome that is called *kwashiorkor* in West Africa. In terms of the minimal model, the phase point enters the region on the singular surface of z_{GK} below the one-dimensional main dynamics manifold, which has been labelled as such in Fig. 6.5.[‡]

In overfeeding, the phase point moves along the right of the one-dimensional manifold, as x_K and x_G both increase in accordance with the compromise that is struck in frustrated homeostasis. The stringency hypothesis, Eq. (6.15), implies that this compromise favours x_G; this is why the z_{GK}-singular surface aligns mostly in the direction of the x_K-axis. Another way of putting this is that the lipostat is less stringent than the glycostat (while both are less stringent than the myostat).

[‡]In kwashiorkor, the lack of selected essential amino acids may be a key factor, particularly those containing biogenic elements not accounted for in the minimal model, and the ability to synthesise new proteins and turn over existing ones may be temporarily impaired or have reduced capacity.

A prominent increase in x_K is labelled *obesity*. When obesity continues to compound due to overfeeding, the capacity for fat storage is attained at some point (the maximum allowed value of x_K, corresponding to the number of adipocytes in the body, each of which can take up and store triglyceride reserves to many times its original volume, but not indefinitely). This results in another sarcopenic crisis. In Fig. 6.5, this has been labelled 'leptin crisis' — rather tentatively, the hormone leptin serving as a token of the fact that adipocytes are filled to capacity. In principle, kwashiorkor and this sarcopenic crisis may occur together, depending on the levels of protein in the diet relative to carbohydrates and fat.

We have seen that along the one-dimensional manifold, there is a section with $x_K < \widehat{x_K}$, characterised by growth cessation, as well as a section with $x_K > \widehat{x_K}$ characterised by growth adjusted to dietary supply and the set-point hierarchy. In the diagrams, the point $x_K = \widehat{x_K}$ has been placed near the center of the diagrams, to facilitate the topological insight into the phase flow. However, in biological reality, one should think of $\widehat{x_K}$ as being located rather closer to the minimum value of x_K.

In adult mammals, both the maximum and minimum values of z_L become zero, as somatic growth (increase of x_L) is no longer possible. As a consequence, the one-dimensional manifold is no longer necessarily attracting in the region where $x_K > \widehat{x_K}$. If nitrogen intake exceeds the obligatory minimum $z_{NX,min}$, the phase point will move upward towards the singular manifold of z_{NX}.

Extensions of the minimal model The main objective of the present case study has been to show how the theory of interlocking control loops yields qualitative insights into the behaviour of the mammalian growth system elucidating why the classic models, which either sport x_L as their sole state variable, or x_L plus an 'energy density' as two state variables, can be accorded validity over a wide range of (re)feeding conditions. We have seen that zero-, one-, or two-dimensional manifolds act as global attractors, depending on diet and food intake history.

Whereas the minimal model is sufficient to demonstrate these points, numerous refinements come to mind. Exploring these at length would venture too far beyond the scope of the present monograph. Let us here touch briefly on the modifications that seem most pressing.

First, the overfeeding crisis at the far end of the obesity branch may be averted if $\widehat{x_K}$ could be adjusted upwards. Physiologically, this is simply a matter of increasing the population of adipocytes [22, 34]. In mathematical modelling terms, we require dynamics for $\widehat{x_K}$ so as to treat it as an $\mathcal{X}\sim$variable (cf. Fig 1.1).

Second, we have treated all $\mathcal{Z}$~variables as fixed and we should allow these to become dynamic. Of particular interest in this respect are z_{KX}, which corresponds to the total energy expenditure (*basal metabolic rate* plus activity, or, equivalently, mechanical work and heat production) and the feeding multiplier z_F, which corresponds to *appetence* [79]. By the general scheme, Eq. (3.21), we already have dynamics, and it remains to introduce appropriate terms in the physiological Lagrangian L.

The feeding multiplier $z_F \in [0, 1]$ is distinct from the food availability *per se* in the environment (which exerts an influence on diet composition), as may be seen from Eq. (6.1). It appears reasonable to assume that L has a minimum at some value $0 < z_F^* < 1$, and that the crisis that occurs when energy reserves are at capacity drives z_F below z_F^*, whereas depletion of the reserves in starvation drives z_F above z_F^*. Evidence for the latter is implicit in the phenomenon of *hyperphagia*, elevated appetence that exceeds the normal feeding drive, up to almost a factor two in rats [82]. Reduction of z_F when the adipocytes are filled to capacity is also well-documented; this involves a hypothalamic pathway whereby the hormone leptin modulates the POMC neurones in the arcuate nucleus [60]. Interestingly, these POMC neurones are also instrumental in modulating the metabolic rate z_{KX} [44]; the food intake ('hunger' and 'satiety') centres are upstream from the TRH-TSH-T3/T4 axis, indicating that, while the system may increase energy expenditure to manage the 'leptin crisis,' this is moderated by food availability and the feeding drive.

It has been suggested that the *energy density* follows simple first-order dynamics in virtually all forms of life [47]; specifically, this density would universally relax to a value governed by the energy density of the diet.[††] In terms of the minimal mammalian model, we should identify this generic energy density with the single coordinate required to specify the position of the phase-point along the one-dimensional manifold. To deal with kwashiorkor-type crises, we would have to supplement this with an auxiliary coordinate, since, as we have seen, the phase point will

[††]This universality has been justified on various grounds, as detailed in ref. 47. Let us sketch, in broad strokes, an interesting meta-selection argument for (nearly) first-order dynamics of nutrient densities. Allowing for a cost-of-growth term $\propto \dot{x}_L x_L^{-3}$ in $\phi_\star$, Eq. (6.2), and imposing first-order dynamics on $n_\star$, we find that $\dot{x}_L$ (growth) depends on $n_\star$ (the 'energy density') according to a rational function. Such functions are *closed under composition* (if f and g are rational, then so is $f \circ g$, up to a caveat that is not relevant here). Food chains, symbiosis, and endosymbiosis have been crucial to the evolution of life on Earth; e.g., eukaryotes evolved from prokaryotes via endosymbiontic embedding; multicellular life arose from acellular organisms by symbiontic aggregation. The evolution of such systems may have been expedited by the fact that bio-energetic 'response characteristics' were preserved as a result of the closure property.

in such cases be (close to) a two-dimensional manifold, and a second coordinate is needed to express, in effect, the degree to which protein-energy malnutrition is disrupting homeostasis. The formalism in hand does not *prima facie* imply (or even suggest) first-order kinetics for the first coordinate. However, we may still exert our freedom in choosing the Lagrangian term for z_F; one may readily see that in this way we can obtain dynamics virtually indistinguishable from first-order relaxation.

Another matter that we have thus far neglected is that the set-points may themselves be subject to dynamic changes. The myostat set-point $\widetilde{x}_{\mathrm{N}}$ represents a degree of muscularity. Both activity patterns and seasonal variation may exert an influence on the degree of muscularity that is most conducive to increased fitness (the next chapter will explore the connection between set-points, L-gradients and evolutionary success at length). A similar seasonal variability may affect the lipostat set-point $\widetilde{x}_{\mathrm{K}}$ which, for the Northern Hemisphere, would suggest that $\widetilde{x}_{\mathrm{K}}$ rises during the autumn and falls during the spring. The phenomenon of homeostatic mechanisms with shifting set-points is known as *allostasis* [18], although the (potentially confusing) term *non-homeostatic regulation* is also in use [37].

Finally, and perhaps most importantly, we have neglected expenditures on reproduction altogether. In principle, since all we really have been doing is the bookkeeping of the three basic species N, $\mathrm{C_G}$, and $\mathrm{C_G}$, it is not difficult to include additional destinations for these atoms, such as offspring. However, a complete treatment is severely complicated by the stoichiometric variability of parental investments, putting the topic firmly beyond the scope of the present monograph.

6.4 EXPLICIT DYNAMICS OF THE MINIMAL MODEL IN THE DIFFERENTIAL INCLUSION-LIMIT

As indicated in Section 5.3, every subregion of the phase space, induced by the DI-partitioning, has its own combination of ODEs and algebraic constraints. Here we set them out explicitly for the minimal mammalian model.

First, when the phase point is in none of the discontinuity surfaces, all relevant $\mathcal{Z}$~variables are assigned either a minimum or a maximum value. For simplicity, we fix z_F, z_{KX} and z_{GX} at constant values and regard z_{NX}, z_L, and z_{GK} as variable. By the argument of Section 5.3, there are $2^3 = 8$ regions in $(x_{\mathrm{N}}, x_{\mathrm{G}}, x_{\mathrm{K}})$-space, in which z_{NX}, z_L, and z_{GK} are set to either their minimum or maximum value. In these regions, the dynamics is free; i.e. Eq. (6.12) applies as stated.

Each of the discontinuity surfaces bounding the regions is characterised by a particular constrained dynamics. To express this explicitly, let us parametrize these surfaces as follows:

$$z_{NX} \text{ singular:} \quad x_N = \xi_N^{NX}(x_G, x_K) \tag{6.16}$$

$$z_L \text{ singular:} \quad x_N = \xi_N^{L}(x_G, x_K) \tag{6.17}$$

$$z_{GK} \text{ singular:} \quad x_G = \xi_G^{GK}(x_K, x_N) \tag{6.18}$$

where the ξs denote functions in two arguments. These constraints yield additional equations for the dynamics, for instance, on the discontinuity surface where z_{NX} is singular, we have $\dot{x}_N = \xi_N^{NX,G}\dot{x}_G + \xi_N^{NX,K}\dot{x}_K$; here $\xi_N^{NX,G}$ is shorthand for the partial derivative $\partial\xi_N^{NX}/\partial x_G$, etc. This can be used to eliminate the dynamics from the free system (6.12) and obtain an expression for the singular value of z_{NX}:

$$\begin{aligned} z_{NX} = {} & \left(1 + (C_{GN} - C_{GK}C_{KN})\xi_N^{NX,G} + C_{KN}\xi_N^{NX,K}\right)^{-1} \\ & \Bigg((u_N - u_G\xi_N^{NX,G} - u_K\xi_N^{NX,K})\frac{z_F}{x_L} + ((1+\kappa C_{GK})\xi_N^{NX,G} - \kappa\xi_N^{NX,K})z_{GK} \\ & + \xi_N^{NX,G}z_{GX} + (\xi_N^{NX,K} - C_{GK}\xi_N^{NX,G})z_{KX} + 3(\xi_N^{NX,G}x_G + \xi_N^{NX,K}x_K - \xi_N^{NX})z_L\Bigg) \end{aligned} \tag{6.19}$$

where the Z~variables on the right are all fixed, so that z_{NX} only depends on x_G and x_K. The two-dimensional dynamics for the latter two state variables, x_G and x_K, is obtained from the free system (6.12) with Eqs. (6.16) and (6.19).

Similarly, on the discontinuity surface where z_L is singular, we have $\dot{x}_N = \xi_N^{L,G}\dot{x}_G + \xi_N^{K,K}\dot{x}_K$ and this can be used to eliminate the dynamics from the free system (6.12) to obtain an expression for the singular value of z_L:

$$\begin{aligned} z_L = {} & \frac{\left(x_N - x_G\xi_N^{L,G} - \xi_N^{L,K}x_K\right)^{-1}}{3}\Bigg((u_N - u_G\xi_N^{L,G} - u_K\xi_N^{L,K})\frac{z_F}{x_L} + \\ & ((1+\kappa C_{GK})\xi_N^{L,G} - \kappa\xi_N^{L,K})z_{GK} + \xi_N^{L,G}z_{GX} - (C_{GK}\xi_N^{L,G} - \xi_N^{L,K})z_{KX} \\ & -(1 + (C_{GN} - C_{GK}C_{KN})\xi_N^{L,G} + C_{KN}\xi_N^{L,K})z_{NX}\Bigg). \end{aligned} \tag{6.20}$$

The two-dimensional dynamics for the free state variables, x_G and x_K, is obtained from the free system (6.12) with Eqs. (6.17) and (6.20).

Finally, on the discontinuity surface where z_{GK} is singular, we have $\dot{x}_{\mathrm{G}} = \xi_{\mathrm{G}}^{\mathrm{GK,K}}\dot{x}_{\mathrm{K}} + \xi_{\mathrm{G}}^{\mathrm{GK,N}}\dot{x}_{\mathrm{N}}$ and this can be used to eliminate the dynamics from the free system (6.12) to find the singular value of z_{GK}:

$$z_{\mathrm{GK}} = \left((u_{\mathrm{G}} - u_{\mathrm{K}}\xi_{\mathrm{G}}^{\mathrm{GK,K}} - u_{\mathrm{N}}\xi_{\mathrm{G}}^{\mathrm{GK,N}})\frac{z_F}{x_L} - z_{\mathrm{GX}} + (C_{\mathrm{GK}} + \xi_{\mathrm{G}}^{\mathrm{GK,K}})z_{\mathrm{KX}} + \right.$$
$$- 3(\xi_{\mathrm{G}}^{\mathrm{GK}} + \xi_{\mathrm{G}}^{\mathrm{GK,K}}x_{\mathrm{K}} + \xi_{\mathrm{G}}^{\mathrm{GK,N}}x_{\mathrm{N}})z_L +$$
$$\left. (C_{\mathrm{GN}} - C_{\mathrm{GK}}C_{\mathrm{KN}} - C_{\mathrm{KN}}\xi_{\mathrm{G}}^{\mathrm{GK,K}} + \xi_{\mathrm{G}}^{\mathrm{GK,N}})z_{\mathrm{NX}} \right)(1 + \kappa C_{\mathrm{GK}} + \kappa\xi_{\mathrm{G}}^{\mathrm{GK,K}})^{-1} \quad (6.21)$$

The two-dimensional dynamics for the free state variables, x_{K} and x_{N}, is obtained from the free system (6.12) with Eqs. (6.18) and (6.21).

The discontinuity surface with singular z_{GK} is the most important one from a physiological point of view, since rather extreme variations in dietary composition must be imposed to take the phase point a substantial distance away from this surface. Within each discontinuity surface, we have discontinuity curves where the surface intersects another one. Given the central role of the surface with singular z_{GK}, we focus on the discontinuity lines on this surface. There are two of these: one where the surface intersects with the surface with singular z_{NX} and one where it intersects with the surface with singular z_L. Of these, the former is important in starvation and the latter in regular growth; only under conditions of severe starvation or during recovery from severe starvation will the phase point lie significantly below these intersection curves. Thus the resulting one-dimensional models (with only x_{K} as a free state variable) are valid simplifications under most eco-physiological circumstances.

On the 'energy deficient' arc (i.e. the left-most part of the one-dimensional manifold depicted in Fig. 6.5), the singular values of z_{GK} and z_{NX} are obtained by eliminating the dynamics from the free system (6.12) using the two conditions $\dot{x}_{\mathrm{G}} = \xi_{\mathrm{G}}^{\mathrm{GK,K}}\dot{x}_{\mathrm{K}} + \xi_{\mathrm{G}}^{\mathrm{GK,N}}\dot{x}_{\mathrm{N}}$ and $\dot{x}_{\mathrm{N}} = \xi_{\mathrm{N}}^{\mathrm{NX,G}}\dot{x}_{\mathrm{G}} + \xi_{\mathrm{N}}^{\mathrm{NX,K}}\dot{x}_{\mathrm{K}}$ and solving for z_{GK} and z_{NX}. Near this discontinuity curve, the partial derivatives of the functions $\xi_{\mathrm{G}}^{\mathrm{GK}}$ and $\xi_{\mathrm{N}}^{\mathrm{NX}}$ are negligibly small and it is thus of interest to display the expressions for the singular $\mathcal{Z}$~variables with these derivatives set to zero:

$$z_{\mathrm{GK}} = \frac{(u_{\mathrm{G}} + (C_{\mathrm{GN}} - C_{\mathrm{GK}}C_{\mathrm{KN}})u_{\mathrm{N}})z_F/x_L - z_{\mathrm{GX}} + C_{\mathrm{GK}}z_{\mathrm{KX}}}{1 + \kappa C_{\mathrm{GK}}} \quad (6.22)$$

$$z_{\mathrm{NX}} = u_{\mathrm{N}}z_F/x_L \quad (6.23)$$

(here $z_L = z_{L,\min} = 0$ due to the topology of the various surfaces). The 'energy surplus' curve (i.e. the right-most part of the one-dimensional manifold depicted in Fig. 6.5) can be treated in similar fashion, resulting in the following expressions for the singular variables:

$$z_{GK} = \frac{(u_G x_N - u_N x_G) z_F / x_L + x_G z_{NX}}{x_N(1+\kappa C_{GK})} + \frac{x_N(C_{GK} z_{KX} - z_{GX} + (C_{GN} - C_{GK} C_{KN}) z_{NX})}{x_N(1+\kappa C_{GK})} \tag{6.24}$$

$$z_L = \frac{u_N z_F / x_L - z_{NX}}{3 x_N} \tag{6.25}$$

(here $z_{NX} = z_{NX,min}$ due to the topology of the various surfaces and $x_N \approx \widehat{x_N}$, $x_G \approx \widehat{x_G}$, for the section of the curve where the ξ-derivatives can be ignored).

Along the 'energy deficient' curve, the ordinary differential equation for x_K is as follows:

$$\dot{x}_K = \left(u_K + \frac{\kappa u_G}{1+\kappa C_{GK}} - \frac{(C_{KN} + \kappa C_{GN}) u_N}{1+\kappa C_{GK}} \right) \frac{z_F}{x_L} - \frac{z_{KX} + \kappa z_{GX}}{1+\kappa C_{GK}} \tag{6.26}$$

whereas along the 'energy surplus' curve, we have:

$$\dot{x}_K = \left(u_K + \frac{\kappa u_G}{1+\kappa C_{GK}} - \frac{(\kappa x_G + (1+\kappa C_{GK}) x_K) u_N}{(1+\kappa C_{GK}) x_N} \right) \frac{z_F}{x_L} - \frac{z_{KX} + \kappa z_{GX}}{1+\kappa C_{GK}} . \tag{6.27}$$

The three discontinuity surfaces intersect in the homeostatic point $\widehat{x_N}$, $\widehat{x_G}, \widehat{x_K}$. To maintain these three state variables at these values, the following singular control conditions must be satisfied:

$$z_L = \frac{(\kappa u_G + (1+\kappa C_{GK}) u_K + (C_{KN} + \kappa C_{GN}) u_N) \frac{z_F}{x_L} - \kappa z_{GX} - z_{KX}}{3(\kappa \widehat{x_G} + (1+\kappa C_{GK}) \widehat{x_K} + (C_{KN} + \kappa C_{GN}) \widehat{x_N})} \tag{6.28}$$

$$\begin{aligned} z_{NX} = {} & \frac{(\widehat{x_K} + \kappa(\widehat{x_G} + C_{GK} \widehat{x_K})) u_N - \kappa \widehat{x_N} u_G - (\widehat{x_N} + C_{GK} \kappa \widehat{x_N}) u_K}{\kappa \widehat{x_G} + (1+\kappa C_{GK}) \widehat{x_K} + (C_{KN} + \kappa C_{GN}) \widehat{x_N}} \frac{z_F}{x_L} \\ & + \frac{\kappa \widehat{x_N} z_{GX} + \widehat{x_N} z_{KX}}{\kappa \widehat{x_G} + (1+\kappa C_{GK}) \widehat{x_K} + (C_{KN} + \kappa C_{GN}) \widehat{x_N}} \end{aligned} \tag{6.29}$$

$$\begin{aligned} z_{GK} = {} & \frac{(C_{GN} \widehat{x_K} - C_{KN}(\widehat{x_G} + C_{GK} \widehat{x_K})) u_N + (\widehat{x_K} + C_{KN} \widehat{x_N}) u_G}{\kappa \widehat{x_G} + (1+\kappa C_{GK}) \widehat{x_K} + (C_{KN} + \kappa C_{GN}) \widehat{x_N}} \frac{z_F}{x_L} \\ & - \frac{(\widehat{x_G} + (C_{GN} - C_{GK} C_{KN}) \widehat{x_N}) u_K}{\kappa \widehat{x_G} + (1+\kappa C_{GK}) \widehat{x_K} + (C_{KN} + \kappa C_{GN}) \widehat{x_N}} \frac{z_F}{x_L} \\ & + \frac{(\widehat{x_G} + C_{GK} \widehat{x_K} + C_{GN} \widehat{x_N}) z_{KX} - \widehat{x_K} z_{GX} - C_{KN} \widehat{x_N} z_{GX}}{\kappa \widehat{x_G} + (1+\kappa C_{GK}) \widehat{x_K} + (C_{KN} + \kappa C_{GN}) \widehat{x_N}} . \end{aligned} \tag{6.30}$$

The classic equation The mode represented by Eqs. (6.24) and (6.25) applies whenever the organism is not subjected to, or recovering from, severe starvation, so that homeostasis can be maintained for both C_G

and N reserves. In this mode, growth is essentially nitrogen-driven (or nitrogen-limited, depending on how one prefers to state such matters); i.e., the increase in lean biomass is directly tied to the dynamic nitrogen balance. As noted in Section 6.1, Eq. (6.25) reduces to the classic Pütter-Bertalanffy model when all terms except x_L are constant in time:

$$\dot{x}_L = z_L x_L = \frac{u_N z_F - z_{NX} x_L}{3x_N} . \tag{6.31}$$

In this context, with homeostasis of C_G and N reserves assured (or assumed), we can take lean biomass to be the total biomass less the adipose reserve mass, the latter being proportional to $r_K x_L^3$ (if variations in water weight are substantial, this calculation is best couched in terms of dry weights).

The spirit of the present approach is not to reject the classic models, but to delineate their domains of applicability in terms of homeostasis for the three basic species N, C_G, and C_G. It is interesting that the first of these species is the primary driver within this domain of applicability — though not exclusively so, since the energy density status will affect basal metabolic rate and feeding multiplier.

Exit conditions Considerations of fundamental physiology restrict each $\mathcal{Z}$~variable to a corresponding range, i.e., $0 \leq z_L \leq z_{L,\max}$, $z_{NX,\min} \leq z_{NX} \leq z_{NX,\max}$, and $z_{GK,\min} \leq z_{GK} \leq z_{GK,\max}$. A discontinuity surface forms the boundary between two regions of phase space in which the $\mathcal{Z}$~variable associated with that surface assumes, respectively, its minimal and maximal allowed values. One thus has dynamic flows on either side of the surface, and for a given point on the surface one may evaluate the components of these two flows that are normal to the surface. Thus, for the sake of clarity, let us say that we have a flow on the 'min' side of the discontinuity surface and a flow on the 'max' side, and that the components of these flows normal to the surface are labelled the 'min vector' and the 'max vector'. There are several cases to consider.

(i) Stable slide: the min vector is either zero or pointing toward the max-side, *and* the max vector is either zero or pointing toward the min-side. It follows from the continuous dependence of the flow on the $\mathcal{Z}$~variable that for at least one value of z inside the allowed range (boundaries inclusive) the normal component of the flow vanishes. For the system under consideration, this singular value is in fact unique, as is immediate from the explicit formulas in the foregoing equations. Thus the $\mathcal{Z}$~variable assumes this singular value and we have constrained dynamics on the surface.

(ii) Unstable slide: the min vector is either zero or pointing toward the min-side, *and* the max vector is either zero or pointing toward the max-side. For the present system, we can exclude the possibility that both min and max vectors are zero, restricting our attention to scenarios in which the normal component varies as the $\mathcal{Z}$~variable travels through its allowed range. Then we have constrained dynamics on the discontinuity surface, as in case (i), but it is unstable under perturbations. Let us assume that a small amount of noise is imposed. This will induce the phase point to jump on the diverging phase flow toward either the min-side or max-side, depending on the random perturbation.

(iii) Transveral: the min and max vectors are either zero or pointing toward the min-side, *or* the min and max vectors are either zero or pointing toward the max-side. Ruling out the case where both min and max vectors are zero, we see that the dynamics is not constrained to flow along the discontinuity surface but instead traverses the surface in the direction of the side indicated.

If the phase point is on exactly one discontinuity surface and one of the above 'exit' conditions is satisfied, then the original free dynamics is adopted. If the phase point is on two or more discontinuity surfaces, the exit conditions must be evaluated for each of these surfaces. That is, the min and max vectors are calculated for each discontinuity surface, based on the free flow, as before (i.e., as if the other intersecting continuity surfaces were not there). However, the dynamics is the one confined to the intersection of the surfaces that satisfy the conditions associated with case (i) or case (ii).

Caveats 'Hard' non-linear switching between distinct sets of ODEs, whose algebraic appearance is far from elegant, will seem anathema to some. Life scientists may well ask how the control system 'knows' all those stoichiometric coefficients, and seasoned mathematical biologists will be skeptical, mindful of the problems usually associated with patchwork models encumbered by different dynamics on different parts of the state space.

The answer to such objections is that the 'underlying' model, whose dynamics live on the space spanned by $\{x_L, x_\mathrm{N}, x_\mathrm{G}, x_\mathrm{K}, z_L, z_\mathrm{GK}, z_\mathrm{NX}\}$ is smooth, uniform over the entirety of this space, and merely assumes that the control systems receive afferent information encoding the status of $(x_\mathrm{N}, x_\mathrm{G}, x_\mathrm{K})$. This full dynamics is given by Eq. (3.25), with Eq. (6.12) specifying the physiological component and Eq. (6.15) specifying the regulatory component.

The differential inclusion (DI) formalism describes the essential behaviour of this full system on the three-dimensional state space spanned by $\{x_N, x_G, x_K\}$, under suitable limiting conditions that essentially amount to a time-scale separation, as we saw in Chapter 5. The DI system describes *emergent* behaviour in the true sense of that word. The algebraic thickets in the foregoing series of equations are just a superficial manifestation of a stoichiometric matrix being inverted. The best way to think of the minimal model is as the Pütter-Bertalanffy-Kooijman framework, extended and corrected for the conservation law on nitrogen and the uni-directional conservation law on glucogenic and ketogenic carbon. The original Pütter-Bertalanffy model represents the special case where the $\mathcal{Z}$~system is fully singular, with glucostat, myostat, and lipostat all stringently satisfied. The Kooijman extension countenances the relaxation of the lipostat set-point, in accordance with the set-point hierarchy, Eq. (6.15), whilst maintaining the assumption of strict homeostasis for the glycostat and the myostat.

It is only under conditions of severe starvation that the frustration between the glyostat and the myostat becomes sufficiently pronounced that the assumption $x_G \approx \widehat{x_G}$ is no longer warranted, and only when the carbohydrate- and lipid-based reserves are depleted that net breakdown of the muscle mass becomes an important factor in the organismal energy budget. These are the conditions where the other regulatory modes of the model become relevant. In particular z_{NX} no longer resides at its obligatory minimum (i.e., its lower Anschlag) and z_{GK} is no longer exclusively adjusted, in essence to protect the glycostat set-point.[‡‡]

[‡‡]Strictly speaking, what we are calling the 'glycostat set-point' covers both the glycogen reserves and the blood plasma glucose reserves. The glycostat proper refers only to the former, whereas the latter belongs to the *glucostat.*

CHAPTER 7

The evolutionary perspective

We have argued that set-points are emergent properties that arise from the interactions of the components that make up the control loop. These components are variegated and span all levels of organisation, from molecular to systemic, and while it may not always be straightforward to represent a set-point as a function of ontologically less problematic parameters, there is no *a priori* impediment to postulating that such functions exist. It follows that set-points can quite generally be countenanced in biological regulatory systems when viewed as compound parameters, as discussed in Section 2.1.

Moreover, from the optimal-control point of view, set-points are extreme points of the physiological Lagrangian L. The reification of set-points would therefore be supported if L could be endowed with a reality beyond a convenient formalism. Insofar as we have been correct to suppose that hormones (as well as other types of signals arising in biological control) betoken evolutionary pressures, L ought to be a reflection of the fitness landscape. In this chapter we explore arguments in support of this point of view.

In the jargon of mathematics, we seek to establish both necessity and sufficiency, and mathematicians know all too well that either one or the other is usually hard to come by. For the problem at hand, the 'sufficient' clause seems almost self-evident. If the gradients of L agree with that of the fitness landscape, an obvious selective advantage accrues. For natural selection averages over the fully realised developmental paths of numerous individuals, and picks up *ceteris paribus* correlations, for instance, a negative correlation between reproductive success and blood plasma glucose regularly exceeding 20 mM. The restorative pressure of the mammalian insulin feedback loop confers selective advantage precisely because it avails an individual organism of the evolutionary experience of numerous life histories over countless generations. This much is almost self-evident;[†] what is less clear is how the idea of fitness landscape can be made precise in a way that fits both our understanding of evolutionary processes and its role as a model for L.

The 'necessary' clause is less easily established. If one accepts that evolution maximises fitness; that fitness can be expressed as a functional over the organism's lifetime, or a relevant portion of it; that regulatory systems have evolved as fitness maximisation devices; and, finally, that the QSC approximation is accurate — then the identification of L and fitness landscape is immediate. However, this chain of reasoning presents an overly simplistic view of optimisation in evolution, and of the co-evolution of control systems and the components they regulate. Even if we accept the QSC approximation, we would still have to come to terms with the fact that the fitness cost/benefit balance of investment in regulatory infrastructure is not at all straightforward. One has to reckon with economies of scale, the effects of sharing of signalling paths, the phenomenon that co-opting and adjusting existing structures may be the path of least resistance taken by evolution, and so on.

These are not issues we can straighten out within the scope of this monograph. Instead, we proceed on the assumption that some appropriately qualified version of the simplistic argument may be able to demonstrate that the physiological Lagrangian is, at the very least, shaped by the fitness landscape as far as its general topography is concerned. In this chapter, we survey some of the underlying issues, including several current definitions of fitness (Section 7.1), with particular emphasis on a general definition of fitness that serves our purposes best (Section 7.2) and a precise construction of the fitness landscape (Section 7.3), leading up to a sketch of how the connection can be made (Section 7.4).

7.1 DEFINING FITNESS

To relate an objective functional such as J in Eq. (3.1) to fitness, the most obvious connecting quantity would be the expected accumulated lifetime

[†]In the present chapter we will rely heavily on thought experiments; however, the statement on loss of euglycæmia and reduced fitness has empirical support. In particular, female specimens of *Xenopus tropicalis* that were exposed to an endocrine disruptor (benzo[*a*]pyrene or triclosan) developed a diabetes-like metabolic syndrome, accompanied by a decline in reproductive success [73].

fecundity, which can be represented as an integral over age a, where the integrand $\lambda(a)$ represents fecundity at a (conceived here as a rate), multiplied by the probability of survival to a [24]. This integral is the *renewal integral*. Approaching fitness via the renewal integral engenders a host of formidable difficulties that are all surmountable in principle [53] but would necessitate case-by-case *ad hoc* modelling approaches, a line of attack which we do not favour, in view of our interest in general theory.

First, we should be wary of a straightforward, naïve interpretation of fecundity, e.g. in terms of egg production and parental care, because an individual can contribute in various ways to the propensity of the particular alleles present in its own genome to recur in later generations, and such investments are not always immediately recognisable as direct reproductive investments. For instance, care for genetically related individuals (kin) should be accounted for, taking into account the degree of genetic relatedness (kinship), and this in turn requires that the latter be always well-defined. Moreover, some kinds of investment can be quite indirect, via influences exerted on the environment, so that the spatiotemporal structure of the organism's eco-physiological setting becomes relevant.

Second, our attention in this monograph is almost exclusively trained on physiological (sub)systems and the degree to which the regulatory components are able to set limits to the amplitude of fluctuations of the physiological component. The selective advantage of setting such limits would reside in an increase of the chances of survival to age a, as well as an improvement of the flow of energy and matter that can be devoted to the overall investment. Yet if these limits are so close together that these benefits are outweighed by the cost of maintaining the regulatory infrastructure, the selective advantage would be diminished. Accordingly, we anticipate an optimal stringency of homeostatic control, where the cost of regulation balances the advantages of enhanced physiological efficiency.

Third, since we are primarily interested in the evolution of *traits*, any approach that is formulated in terms of population genetics (allele frequencies) would have to rely on some genotype-to-phenotype mapping, a challenge that has been partially met for certain tissue systems [5], but which is usually evaded by resorting to highly simplified, schematic functions [12].

It is generally accepted that fitness can be expressed in terms of the *intrinsic rate of increase* r which satisfies

$$1 = \int_0^{\infty} \exp\{-ra\}\lambda(a)da \tag{7.1}$$

(ref. 24). If the average age at which individuals produce offspring is $\bar{a}$, then $r \approx \ln\{\int_0^\infty \lambda(a)da\}/\bar{a}$, an approximation which becomes better as offspring production is more concentrated at age $\bar{a}$ [56]. This formula lends support to the habitual identification of fitness and renewal integral in the evolutionary literature.

The use of intrinsic rate of increase r as a measure for fitness might seem to presuppose that positive fitness can only exist in the context of Malthusian (i.e., exponential) population growth. One can readily show that fitness does not necessarily equal the instantaneous rate of change of ln{biomass} [15, 62], notwithstanding the widespread tradition in microbiology of equating the two [51, 70, 75]. Natural selection happens in populations that are stationary or even shrinking.

Example 20: simple type substitution. Consider two clonally reproducing types, A and A′. Let N denote the fixed population size. In the i generation, $f_{\mathrm{A}}(i)$ individuals are of type A and $f_{\mathrm{A}'}(i) = 1 - f_{\mathrm{A}}(i)$ individuals are of type A′. Type A individuals together produce $f_{\mathrm{A}}(i)\mathrm{N}r_{\mathrm{A}}$ spores, where r_{A} is a fixed parameter. Similarly, type A′ individuals produce $f_{\mathrm{A}'}(i)\mathrm{N}r_{\mathrm{A}'}$ spores, where $r_{\mathrm{A}'}$ is a fixed parameter. The probability that an A′-type spore becomes a reproductively active adult is given *pro rata*, that is, this probability equals $f_{\mathrm{A}}(i)\mathrm{N}r_{\mathrm{A}}/(f_{\mathrm{A}}(i)\mathrm{N}r_{\mathrm{A}} + f_{\mathrm{A}'}(i)\mathrm{N}r_{\mathrm{A}'})$. Provided that N is large enough to allow reasonable appeal to the law of large numbers, we can take this probability to equal $f_{\mathrm{a}}(i+1)$. This gives the recursion:

$$f_{\mathrm{A}}(i+1) = \frac{f_{\mathrm{A}}(i)}{f_{\mathrm{A}}(i) + r_{\mathrm{A}'/\mathrm{A}}(1 - f_{\mathrm{A}}(i))} \tag{7.2}$$

where $r_{\mathrm{A}'/\mathrm{A}} = r_{\mathrm{A}'}/r_{\mathrm{A}}$. We have $\lim_{i\to\infty} f_{\mathrm{A}}(i) = 1$ iff $r_{\mathrm{A}'/\mathrm{A}} < 1$. If this condition is fulfilled and type A is initially rare (i.e., $f_{\mathrm{A}}(i) \ll 1$), a graph of $f_{\mathrm{A}}(i)$ as a function of i will have a sigmoid appearance, charting the time course over trans-generational time of type A starting out as *invader* (or *mutant*) and displacing the *resident* A′. This can be described as a *take-over* event, or also as a A′ → A *substitution*. The number of generations over which this event takes place scales as $(1 - r_{\mathrm{A}'}/r_{\mathrm{A}})^{-1}$ [12]. ❖

It must be admitted that this example is extremely basic (although, in a sense, it contains the essence of natural selection). For instance, more realistic calculations would take into account that r_{A} and $r_{\mathrm{A}'}$ are generally dependent on f_{A}; such *density-dependence* leads to a host of interesting and important phenomena [12]. Nevertheless, the example does demonstrate the argument for an effective uncoupling between the fraction of the focal type, which is the actual quantity of evolutionary interest, and the population size N which may be stationary, or even vary from generation to generation in whatever way circumstances dictate.[‡] It is evident from Example 20 that the take-over dynamics is governed by the *relative fitness advantage*, here expressed by the compound parameter $r_{A'/A}$.

There are several ways of generalising from Example 20. One approach is to focus on the limit $f_A(0) \to 0$. In this limit, the geometric approximation $f_A(i) \approx f_A(0)\left(r_{A'/A}\right)^{-i}$ becomes exact. This case is known as the *infinite dilution Ansatz* [56, 57], because the invader/mutant A is present in a numerical strength that is arbitrarily small compared to the population (of size N), which is composed of one or more resident types.

Moreover, on the infinite dilution Ansatz, each type-A individual only interacts with resident-type individuals. Thus, if there are density-dependent effects, these do not (as yet) come into play. Geometric growth (equivalent to exponential growth in a continuous-time treatment) has the obvious consequence that the focal type (here: A) will not forever stay diluted. However, certain key questions can be answered on the basis of the infinite dilution Ansatz, for instance: how close is the focal type to going extinct, that is, being eliminated by natural selection? What is the probability that the focal type will become established, that is, take over, or become co-existent with one or more residents? If many mutants at infinite dilution abound, which ones will eventually dominate?

To arrive at a proper general concept of fitness, we must refrain from assumptions on non-overlapping, discrete generations, which naturally induce a discrete-time scale. Thus we assume that we are working in continuous time t and that the numerical strength of the focal type is expressed by a time-dependent function $n_A(t)$. We then define the *invasion fitness* of type A as follows:

$$\rho_{A,\boldsymbol{E}} = \lim_{t\to\infty} \ln\{n_A(t)\}/t \tag{7.3}$$

where $\boldsymbol{E}$ denotes the environment, comprising abiotic factors as well as biotic factors, the latter being further divided into conspecific biotic factors (i.e. the resident types which may or may not be ultimately replaced by type A) and other organisms that occur in the ecosystem.

Before delving into the subtleties lurking beneath Eq. (7.3), let us note how knowledge of $\rho_{A,\boldsymbol{E}}$ answers the above questions. Extinction is assured when $\rho_{A,\boldsymbol{E}} \leq 0$; the probability of establishment is positive iff

[‡]In the case where the series $N_0, N_1, N_2, \ldots$ is more generally allowed to be non-constant, a dynamic law is required which may well depend on the distribution of types within the population. The uncoupling only acts in one direction. The argument hinges on the law of large numbers; if this is not applicable, because the population is too small, a more delicate analysis is called for [12].

$\rho_{A,E} > 0$; among many mutants at infinite dilution, the one that maximises $\rho_{A,E}$ will prevail. The biologist's intuition that fitness is precisely what is maximised by natural selection thus finds a natural expression in invasion fitness, although we are left to wonder whether the intrinsic rate of increase r serves this purpose equally well [24, 56].

We can be assured that $\rho_{A,E}$ is a definite real number, if we allow a few additional assumptions. First, the type-A individuals must be statistically independent. This is achieved by *structuring* the population. Structuring variables include age a as well as physiological state variables of the variety we have frequently encountered in this monograph (both $\mathcal{X}$~ and $\mathcal{Z}$~variables), location within the ecosystem, and so on. Generally, these structuring variables are called *h-state variables*, where 'h' stands for 'heterogeneity' [56]. The total numerical strength $n_A(t)$ is obtained by integrating or summing over all possible h-states. If the abiotic part of the environment is possibly chaotic or stochastic but devoid of trends or catastrophic shifts, and there is sufficient h-state structuring to render the type-A individuals statistically independent, then the environment $\boldsymbol{E}$ can be treated as *ergodic*, and then appeal to the multiplicative ergodic theorem can be made [56]. This reifies the invasion fitness $\rho_{A,E}$, viz. the dominant Lyapunov exponent of the h-structured system.

These assumptions, while they may appear restrictive, are on the whole biologically innocuous, and invasion fitness amply fulfils the desiderata for a general fitness concept [56]. The delicate interplay between the two limits required to attain (i) infinite dilution and (ii) the dominant Lyapunov exponent is necessary to achieve the conditional statistical independence on the h-states and the equilibration of the distribution over the h-states, again respectively.[††]

One drawback of the invasion fitness is that the infinite dilution Ansatz confines it to the initial stages of a putative take-over event. However, the very last stages are susceptible to the same Ansatz, but now treating the newly established invader as resident and the nearly extinct former resident as infinitely diluted. Combining the calculations on the initial and final limbs, and observing that together they account for most of the duration of the take-over event, results about the entire transient of the event can be inferred using only the invasion fitness [56].

Another way to generalise Example 20 is to attempt to observe the relative fitness advantage directly over the time course of a take-over

[††] In general, the h-state distribution will be of the mixed discrete/continuous type, but the continuous axes can of course be discretised to any degree of fine-graining so as not to make a difference as far as the biological interpretation is concerned. The point is germane since, strictly speaking, we can only invoke the multiplicative ergodic theorem for discrete h-state distributions, the analogous result for continuous distributions being conjectural.

event. This leads to the concept of *empirical relative fitness*, derived in the following section.

7.2 EMPIRICAL RELATIVE FITNESS

We fix a population at time t_0 and decide upon a binary partition of the extant genotypes in the population into two subsets $\mathcal{K}$ and $\overline{\mathcal{K}}$; that is to say, $\mathcal{K}$ and $\overline{\mathcal{K}}$ cover the genotypes in an exhaustive and mutually exclusive manner. Let $f_{\mathcal{K}}(t)$ represent the fraction of the genotypes belonging to class $\mathcal{K}$ at some point of time $t \geq t_0$ (the time axis is here understood to be on the trans- or even ultra-generational scale, as opposed to t in foregoing chapters, where it denoted 'physiological time,' i.e. a scale comparable with the organism's lifetime or shorter). The 'type' we may associate with $\mathcal{K}$ is said to *take over* and *become fixed* if $\lim_{\tau\to\infty} f_{\mathcal{K}}(t_0+\tau) = 1$ and it is said to be *eliminated by natural selection* if $\lim_{\tau\to\infty} f_{\mathcal{K}}(t_0+\tau) = 0$.[‡‡]

Suppose that the population is finite and allowed to persist for an arbitrarily long time (of course every real biological population is finite both in size and lifetime). Then either one of the outcomes $\lim_{\tau\to\infty} f_{\mathcal{K}}(t_0+\tau) \in \{0,1\}$ is bound to occur by random drift. This will be the case even if there is no fitness differential between $\mathcal{K}$ and $\overline{\mathcal{K}}$. We consider an ensemble of instantiations of the population at time t_0 (a 'heat bath' of infinitely many reference copies) and let $\langle f_{\mathcal{K}}(t_0+\tau)\rangle$ denote the average over this ensemble when it is inspected τ time units later.

Now let

$$\varrho_{\mathcal{K}}(t) = \lim_{\tau\to 0} \frac{\ln\langle f_{\mathcal{K}}(t+\tau)\rangle - \ln\langle f_{\mathcal{K}}(t)\rangle}{\tau} \tag{7.4}$$

$$\sigma_{\mathcal{K}}(t_0) = \lim_{\tau\to 0} \frac{\varrho_{\mathcal{K}}(t_0+\tau) - \varrho_{\mathcal{K}}(t_0)}{\langle f_{\mathcal{K}}(t_0+\tau)\rangle - \langle f_{\mathcal{K}}(t_0)\rangle} \tag{7.5}$$

and consider the Legendre transform

$$\varrho^*_{\mathcal{K}}(t_0) = \varrho_{\mathcal{K}}(t_0) - \sigma_{\mathcal{K}}(t_0)\langle f_{\mathcal{K}}(t_0)\rangle \,. \tag{7.6}$$

[‡‡] These extreme cases are not the only possible outcomes. For instance, *polymorphisms* in which the two types continue to co-exist indefinitely may also occur. The polymorphisms may be genuine, i.e., indefinitely protected, or they may be a transient phase until extinction of one of the two types finally occurs. However, in the latter case, the typical time to final extinction may be much longer than the period over which the population *as a whole* manages to persist [12].

This quantity $\varrho^*_{\mathcal{K}}(t_0)$ defines the *fitness rate coefficient* of class $\mathcal{K}$, relative to the complement class $\overline{\mathcal{K}}$. We have $\varrho^*_{\mathcal{K}}(t_0) > 0$ if and only if $\mathcal{K}$ is evolutionarily favoured relative to $\overline{\mathcal{K}}$.

The special case corresponding directly to Example 20 arises when $\varrho^*_{\mathcal{K}}(t_0 + \tau) \equiv \varrho^*_{\mathcal{K}}(t_0)\ \forall \tau \geq 0$. In this case we have

$$\langle f_{\mathcal{K}}(t_0 + \tau) \rangle = \left(1 - \left(1 - \langle f_{\mathcal{K}}(t_0) \rangle^{-1}\right) \exp\{-\varrho^*_{\mathcal{K}}(t_0)\tau\}\right)^{-1} \tag{7.7}$$

(simple logistic take-over). Another important special case arises when $\varrho^*_{\mathcal{K}}(t_0 + \tau)$ does depend on τ, but *exclusively through* $\langle f_{\mathcal{K}}(t_0 + \tau) \rangle$; this case is referred to as *density dependence*.

The thought experiment involving an ensemble of histories takes a different approach as compared to the one motivating the concept of invasion fitness. The present scenario has the advantage that it can be set up in practice, albeit imperfectly: instead of an ensemble in the true statistical sense, we will have to be satisfied with a finite number of reproductions and we can never be sure that the starting configuration is identical. Nonetheless, certain evolutionary scenarios do allow for some form of replication of the take-over event, for instance, within-host viral evolution, or bacterial/bacteriophage evolution in continuous culture. In such systems, it is possible to obtain the data sets that allow us to carry out the construction of Eqs. (7.4)–(7.6) via numerics. The present construction is also well-suited for *in silico* evolution, where we may study an allele that has different effects on fecundity, depending on the genomic context of the allele (e.g., mating type), where we can build in stochastic and finite population effects and analyse the output of repeated 'runs' of the evolutionary process to generate data that can be processed via Eqs. (7.4)–(7.6); several examples are given in ref. 12.

Although we put no *a priori* constraints on the choice of the subset $\mathcal{K}$ of genotypes, our intuitive paradigm case is where $\mathcal{K}$ contains a particular allele on a locus of interest, and $\overline{\mathcal{K}}$ everything else. However, more complex genotypic types can be constructed through the expedient of Boolean connectives — to the point where $\mathcal{K}$ contains just a single genotype (which could be one particular individual or perhaps a set of identical twins or clones).

Although we have introduced $\mathcal{K}$ and $\overline{\mathcal{K}}$ as genotypic classes, our primary interest is in the evolution of traits, which is essentially why we have eschewed the approach via the renewal integral. Accordingly, we shall assume that we can let $\mathcal{K}$ and $\overline{\mathcal{K}}$ refer to phenotypic classes and obtain empirical relative fitness measures for such classes as well.

The concept of *trait* seems to be rather flexible.[§] We should therefore be careful to restrict our attention to those phenotypic classes such that the associated $\varrho^*_{\mathcal{K}}(t_0)$ is not infinitesimally small, that is to say, the choice of $\mathcal{K}$ has to be *evolutionarily relevant.*

7.3 THE FITNESS LANDSCAPE

Let us assume that it is possible, in principle, to characterise the fully realised developmental path over the lifetime of the individual organism as a function of h-state $\xi(a) \in \mathcal{H}$. Here the h-state space $\mathcal{H}$ is a compact subset of $\mathbb{R}^{n_h}$, $n_h \geq 1$ (and in fact we imagine that $n_h \ggg 1$ for a 'complete' h-state description), and a is the age of the organism, measured from a reference point that seems sensible for the life cycle at hand.

Approaching the problem naïvely, we should like to take the class $\mathcal{K}$ to be all individuals whose life path is given by some specific choice of $\xi(\cdot)$. Unfortunately, any particular life path is far too specific to be evolutionarily relevant (this difficulty is measure-theoretical in nature). The state value itself is a more promising candidate, since the state is, in essence, an equivalence class of system histories [64]. Thus subsets of the state space represent bundles of life histories. Accordingly, judicious choices of such subsets should be evolutionarily relevant. We seek to attach fitness values to state values, and this mapping will be our formal version of what we have hitherto referred to in intuitive terms as the *fitness landscape*.

Let us fix a number $s \in (0, 1]$ and let $\mathcal{X}_s$ denote a compact subset of $\mathcal{X}$ such that $|\mathcal{X}_s|/|\mathcal{X}| = s$. By $\varrho^*_{\xi(a)\in\mathcal{X}_s}(t_0)$ we mean the relative fitness associated with the class of individuals such that their state is in $\mathcal{X}_s$ at age a, at time point t_0 in evolutionary history (thus t_0 generally belongs to a slower, possibly much slower, time scale than a). Our attention is confined to the evolutionarily relevant cases among the subclasses of size (volume) $s|\mathcal{X}|$. Since, for the sake of the general argument, we are

[§]Conventional usage among life scientists seems to defy any attempt at a crisp definition demarcating what constitutes a *trait*: roughly speaking, a trait seems to be any property or characteristic of an organism that can be supposed to exhibit significant correlation with some genotypic property, although we must allow that such correlation involves developmental plasticity and the correlation measure must be appropriately pre-conditioned on e.g. environmental influences. Classically, traits were mostly of the morphological/anatomical variety whilst genes were shrouded in mystery, making for a marked contrast between phenotype and genotype. The advent of biochemical assaying and affordable extensive genotyping has blurred the distinction. For if correlation with some genetic basis were to be the main criterion, the actual presence of a certain DNA sequence at a defined position in the genome ought to count as a 'trait' *par excellence*.

allowing $\mathcal{H}$ to be a biologically exhaustive h-space, this assumption is warranted. Let us write $\varrho^*[\mathcal{X}_s, a]$ for $\varrho^*_{\xi(a)\in\mathcal{X}_s}(t_0)$ and in so doing suppress the dependence on point of time in evolutionary history t_0 (to avoid notational explosion, we display explicitly only what is germane to the specific argument at hand).[¶]

We next consider a state space subset $\widehat{\mathcal{X}}_{s,a}$ such that

$$\varrho^*[\widehat{\mathcal{X}}_{s,a}, a] \geq \varrho^*[\mathcal{X}_s, a]$$

for all state space subsets $\mathcal{X}_s$ of size (volume) $s|\mathcal{X}|$. We assume that $\widehat{\mathcal{X}}_{s,a}$ is unique. For $\boldsymbol{x} \in \mathcal{X}$ we define:

$$\varrho^*(\boldsymbol{x}, s, a) = \begin{cases} \varrho^*[\widehat{\mathcal{X}}_{s,a}, a] & \text{if } \boldsymbol{x} \in \widehat{\mathcal{X}_s} \\ \varrho^*[\mathcal{X} \setminus \widehat{\mathcal{X}}_{s,a}, a] & \text{if } \boldsymbol{x} \in \mathcal{X} \setminus \widehat{\mathcal{X}_s} \end{cases} . \tag{7.8}$$

We can now define an upper bound to the relative fitness of individuals such that $\boldsymbol{\xi}(a) = \boldsymbol{x}$:

$$\varrho^*(\boldsymbol{x}, a) = \sup_{s\in(0,1]} \{\varrho^*(\boldsymbol{x}, s, a)\} \tag{7.9}$$

Suppose that the organism dies at age $a_\dagger$.[♯] Then $\varrho^*(\boldsymbol{x}, a_\dagger)$ denotes the fitness of that organism, considered as a member of a type class defined by dying at age $a_\dagger$ in state $\boldsymbol{x}$.

It is the abstract conception of a state as a bundle (equivalence class) of histories that justifies the assignment of a fitness value to a state.

[¶]Dropping t_0 from our notation, we sweep a pressing problem in evolutionary theory under the carpet. The actual lives that have shaped the genome and developmental biology of an organism lie in the recent evolutionary past, which Godfrey-Smith called the *modern history* [27], which we should view not as a definite segment of recent time (e.g., from a million years ago till now) but rather as a kind of discounted past, with lives lived longer ago generally counting for less (with a discounting rate that might well be different for different traits in the same organism). If typical life histories and environmental conditions are much the same over this 'modern history' period, ϱ^*'s dependence on t_0 can be ignored with impunity. However, if this condition is not fulfilled, there is a *fitness lag*: it may well happen that the organism's systems act in a way that is (virtually) 'fitness-optimal' relative to the fitness ϱ^* at some time in the recent evolutionary past, but is no longer so at present. The significance of fitness lag is far from academic in the present epoch of rapid anthropogenic environmental change. For instance, our central thesis that the endocrine system is shaped by natural selection (via the effects that endocrine signals have on physiology, and the latter on reproductive success) seems to be gainsaid by the finding that acute ozone-induced pulmonary and systemic metabolic effects are diminished in adrenalectomised rats — that is, cutting out an endocrine loop seems to improve health and hence reproductive success [59].

[♯]The type of organism (unitary, modular) and its life cycle determine how the point of death should be ascertained. In general, we mean to denote by $a_\dagger$ the point in time at which the individual, however defined, ceases to be a living entity.

Moreover, if this state is the final state of the life history path, we may claim outright that $\varrho^*(\boldsymbol{x}, a_\dagger)$ denotes the fitness of an individual dying at age $a_\dagger$ in state $\boldsymbol{x}$. However, in the case of a younger age $a < a_\dagger$, we cannot state that $\varrho^*(\boldsymbol{x}, a_\dagger)$ denotes the fitness of an organism considered as a member of a type class defined by being in state $\boldsymbol{x}$ at age a, unless the difference $(a_\dagger - a)$ happens to be infinitesimally small. Rather, it is the best fitness that such an individual, considered as a member of the above type class, could attain, depending on how it lives out the remainder of its life. This follows because, first, whenever $s_1 < s_2$, we have $\widehat{X}_{s_1,a} \subset \widehat{X}_{s_2,a}$ and, second, we take the supremum in Eq. (7.9).

Taken together, the foregoing considerations prompt us to regard fitness as a functional over the life path $\boldsymbol{\xi}(\cdot)$:

$$\varrho^*(\boldsymbol{\xi}(a_\dagger), a_\dagger) = \varrho^*(\boldsymbol{\xi}(0), 0) + \int_{t_*}^{t_\dagger} \left(\nabla_{\boldsymbol{x}} \varrho^*(\boldsymbol{\xi}(\tau - t_*), \tau - t_*) \cdot \boldsymbol{\xi}'(\tau - t_*) + \frac{\partial \varrho^*(\boldsymbol{\xi}(\tau - t_*), \tau - t_*)}{\partial a} \right) d\tau \tag{7.10}$$

where t_* is the time of birth and $t_\dagger$ is the time of death, so that age $a = t - t_*$. Its daunting appearance notwithstanding, the integrand is simply the result of calculating the total differential and observing that $da/dt = 1$.

Let us now suppose, in addition, that we know the dynamics $\dot{\boldsymbol{x}}(t) = \boldsymbol{f}(\boldsymbol{x}(t), \boldsymbol{u}(t))$ where $\boldsymbol{u}(t)$ is the external or environmental input (forcing function).[♭] We do not need to let this dynamics depend explicitly on age a, because age is already encoded in the state $\boldsymbol{x}$. With the dynamics in hand, we rewrite the integral as follows:

$$\int_{t_*}^{t_\dagger} \left(\nabla_{\boldsymbol{x}} \varrho^*(\boldsymbol{x}(\tau), \tau - t_*) \cdot \boldsymbol{f}(\boldsymbol{x}(\tau), \boldsymbol{u}(\tau)) + \frac{\partial \varrho^*(\boldsymbol{x}(\tau), \tau - t_*)}{\partial a} \right) d\tau .$$

The import of this is that the integrand is a function of the state $\boldsymbol{x}$ and the input (ambient forcing) $\boldsymbol{u}$ only. In particular, let us give the integrand its own symbol:

$$\Lambda_{\mathcal{H}}(\boldsymbol{x}, \boldsymbol{u}) = \nabla_{\boldsymbol{x}} \varrho^*(\boldsymbol{x}, \alpha(\boldsymbol{x})) \cdot \boldsymbol{f}(\boldsymbol{x}, \boldsymbol{u}) + \frac{\partial \varrho^*(\boldsymbol{x}, \alpha(\boldsymbol{x}))}{\partial a} \tag{7.11}$$

[♭] It is only by way of thought experiment that we imagine ourselves to dispose over a comprehensive and accurate mathematical representation of *everything* that occurs in a living system over its lifetime, e.g. involving an astronomical number of ordinary differential equations. Putting this in practice could be the great promise of computational biology, or else the *nec plus ultra* of positivistic naïveté. The latter seems more likely, but in any case, to obtain Eq. (7.10), we need only invoke the state concept *in abstracto* to make plausible that fitness can be represented as a functional whose integrand involves no more than state and input (in the system-theoretic senses of these terms).

where we stipulate that the h-state allows the recovery of age from the state $\boldsymbol{x}$, the function $\alpha(\cdot)$ making this link explicit. The function $\Lambda_{\mathcal{H}}(\cdot,\cdot)$ is the fitness landscape. The construction works equally well if we use any monotone function of ϱ^* instead of ϱ^* itself, for example $\exp\{\varrho^*\overline{a}\}$, which connects ϱ^* to the renewal integral. In fact, if conditions are fulfilled such that the intrinsic rate of increase r is identifiable with relative fitness,[||] we have $\Lambda_{\mathcal{H}}(\boldsymbol{\xi}(a),\boldsymbol{u}(a)) \equiv \lambda(a)$ (the pair $(\boldsymbol{\xi}(a),\boldsymbol{u}(a))$ satisfying $\boldsymbol{\xi}'(a) = \boldsymbol{f}(\boldsymbol{\xi}(a),\boldsymbol{u}(a))$).

7.4 THE PHYSIOLOGICAL LAGRANGIAN AS FITNESS LANDSCAPE

Having lined up our chessmen, we will now attempt a direct linking of L and $\lambda_{\mathcal{H}}$. In previous chapters, we have isolated spaces of state variables, $\mathcal{X}$ and $\mathcal{Z}$; these are to be supposed as subspaces of the complete h-space $\mathcal{H}$. Mathematically, we are dealing with a projection from the complete state $\boldsymbol{\xi}$ to the state of immediate interest, $(\boldsymbol{x},z)$; here we are using $\boldsymbol{\xi}$ to denote the complete h-state, as $\boldsymbol{x}$ shall specifically denote an element of $\mathcal{X}$. It is no longer guaranteed that age a is encoded by the pair $(\boldsymbol{x},z)$, and thus we should speak of a projection from $\boldsymbol{\xi}$ to $(\boldsymbol{x},z,a)$.

Under this projection, it may be seen that the argument of Section 7.3 carries through, with certain provisos. Fitness should now be understood as *marginal fitness*, and, moreover, evolutionary relevance is no longer guaranteed, as we should expect, since not all physiological fluctuations are such that diminishing their amplitude would boost fitness. Moreover, we can think of this projection as either motivated by our interest, that is, we happen to focus on a subsystem S that is arbitrarily defined by our perception, or motivated by what the regulatory component can *monitor*, that is, the capability of the afferent pathways to conduct signals conveying information about certain aspects of the h-state $\boldsymbol{\xi}$. In the first case, the delineation of the problem is not objective, strictly speaking, whereas in the second case, this delineation is itself subject to evolution.

Let us consider whether we can obtain a fitness landscape pertaining to the subsystem S at hand, which we write as $\Lambda_S(\boldsymbol{x},z,\boldsymbol{u},a)$ or simply $\Lambda(\boldsymbol{x},z,\boldsymbol{u},a)$ if the context makes clear which subsystem is the focus of attention. As we observed at the beginning of this chapter, the 'sufficient'

[||] We shall not delve into the precise nature of these conditions. They broadly correspond to the conditions that reconcile the renewal integral approach to the invasion fitness approach, as detailed in ref. 56.

part of our thesis is straightforward, and appears to be merely a matter of putting

$$L = -\Lambda$$

where the minus sign appears because we have consistently talked about minimising the functional J (cf. Eq. (3.1)). Three issues need to be addressed. First, the construction in Section 7.3 only yields Λ up to a monotone increasing function on ϱ^*, as we noted above. However, some functions are better choices than others; in particular, applying a transformation of the form $\exp\{\varrho^* T\}$ where T is an appropriate time scale (e.g. mean life cycle time) has a stabilising effect on the ensuing calculations.

Second, an explicit dependence on $\boldsymbol{u}(t)$ remains in Λ. From the point of view of classic homeostasis (maintenance of *milieu intérieur*) the term with the dynamics $\boldsymbol{f}$ in Eq. (7.11) is less important and the dependence on $\boldsymbol{u}(t)$ can be suppressed. More generally, we should allow the input to figure in the physiological Lagrangian. Since we are dealing with organismal subsystems, the input should be expected to be a combination of an environmental (ambient) input and effects arising elsewhere in the organism, and hence properly speaking functions of other h-state variables.

Third, we have an explicit dependence on age a which did not figure in the Lagrangian L as introduced in Eq. (3.1). The dependence on a is crucial for life history trade-offs that involve sequencing of life stages, e.g. the timing and nature of transitioning from larva to pupa, from juvenile to adult, or from a reproductively inactive to a reproductively active stage. These transitions are encoded by the complete fitness landscape $\Lambda_{\mathcal{H}}$ as shifts in $\mathcal{H}$-space maxima along the a-axis. Surges of highly pleiotropic hormones are in keeping with this picture, as these a-dependent shifts suddenly create steep gradients in $\mathcal{H}$-space.

The present theory accounts for hormones as tokens-of-life-history-trade-offs as a special (and rather dramatic) case of hormones-as-tokens-of-evolutionary-pressures (gradients of $\Lambda_{\mathcal{H}}$). However, we have mostly confined our attention to classic homeostasis, not the least because this has allowed us to steer clear of the semantic debate concerning the concept of trade-off [24, 43, 101], which would have needlessly muddled the discussion.

Even from the perspective of classic homeostasis, age a plays a role. In particular, the maxima of Λ tend to sag, that is, both broaden and become less tall, as a approaches the region where the probability mass of $a_\dagger$ is concentrated. This observation essentially restates Hamilton's evolutionary account of senescence [33]. Unless senescence is explicitly the focus of our investigation, we shall usually want the time limits t_0 and t_1 in Eq. (3.1) to span a range of ages over which this sagging is not too prominent, and simply substitute the average age into Λ. There may be additional considerations, e.g., relevant life stages, which would further constrain the choice of t_0 and t_1.

As we have discussed in the introduction to this chapter, the 'necessary' part of our thesis is rather more problematic, for several reasons we have already set out. Nonetheless, let us at least sketch an argument. To simplify matters, let us consider the complete landscape function at a fixed age a^*. Suppose that a fitness investment of magnitude $T\delta\Lambda_{\mathcal{H}} > 0$ is made to support the development, maintenance, and operation of a piece of regulatory machinery. Here T is the typical portion of the life span over which this machinery operates (that is, $T \approx a_\dagger$ for most of the classic homeostasis scenarios we have been discussing in this monograph). Let us furthermore suppose that the machinery is maximally successful at h-state value $\boldsymbol{\xi} \in \mathcal{H}$ in the sense that the presence of the machinery results in a correction $\delta\boldsymbol{\xi}$ in the direction of the gradient $\nabla_\xi \Lambda_{\mathcal{H}}(\xi, a^*)$. The net impact on fitness is favourable if and only if

$$\frac{|\delta\boldsymbol{\xi}|}{\delta\Lambda_{\mathcal{H}}} > \left|\nabla_\xi \Lambda_{\mathcal{H}}(\xi, a^*)\right| \tag{7.12}$$

provided that the δs are small enough to warrant a first-order approximation. With equality, $|\delta\boldsymbol{\xi}|/\delta\Lambda_{\mathcal{H}} = \left|\nabla_\xi \Lambda_{\mathcal{H}}(\xi, a^*)\right|$, the regulatory innovation is selectively neutral. However, even for favourable substitutions we should expect *near*-equality on the assumption that successful mutations will in most cases not substantially exceed the minimum set here. Combining this with the fact that the gradient of L is what generates regulatory intervention in the dynamics as described in Chapter 3, we find that it is eminently reasonable to expect $L \propto \Lambda_{\mathcal{H}}$.

The foregoing argument lacks sophistication. Various modifications and improvements readily come to mind. However, the general idea is already conveyed quite well by the crude sketch in the preceding paragraph. And in any case, the skeptic is hardly likely to be swayed by any refinements as long as our calculations depart from the QSC approximation, or the gradient-driven dynamics that generalise it.

An alternative approach to support the $L \propto \Lambda_{\mathcal{H}}$ hypothesis might be to treat the input $\boldsymbol{u}(t)$ as a random process, so that $\boldsymbol{x}(t)$ also becomes a random process, even if we retain f as deterministic. The behaviour of $\boldsymbol{x}(t)$ can be viewed as a Markov process under a suitable coarse-graining of $\mathcal{X}$, and the tools of potential theory should then allow us to define the fitness landscape. In this framework, the regulatory component may be modelled as 'tilting' the law of this Markov process.

CHAPTER 8

Critique and outlook

A minimal claim for the theory set out in this monograph is that it presents a general recipe for building mathematical models of biological systems. Specifically, those aspects that can be firmly based in well-established principles are explicitly represented in the dynamics f and the remainder is represented by encoding physiological intuitions about behavioural objectives in a function L. The former coincides more or less with the 'peripheral' or 'physiological' component of the system, and the latter with the regulatory mechanisms. Thus the approach is particularly suited to analyse systems where the full architecture of the control system has not been fully charted [14]. Furthermore, as knowledge regarding the regulatory infrastructure becomes available, this can be 'siphoned' from L to f, adjusting both accordingly.

We have introduced the QSC approximation, L-gradient driven dynamics, coupling graphs, and the differential inclusion limit. These tools provide a rationalisation of the pleiotropic suites associated with hormones (as well as, possibly, other first messengers) and of the qualitative phenomena that emerge when control loops interlock and control 'goals' conflict. No claim of novelty is made here for any of these tools, other than the modest hope that they have been brought together in a coherent fashion that elucidates how they interconnect.

❧

Our more expansive claim is more controversial: hormones are tokens of evolutionary pressures and the relationship between fitness and physiology leads to gradient-driven dynamics. On this view, the $L = -\Lambda$ correspondence reifies set-points as the extrema of the fitness landscape. The set-point concept loses its teleological burden, which looms large because engineers freely avail themselves of terms such as 'reference,' 'desired,' 'target,' 'goal,' etc., with the concomitant metaphysical peril in the eyes of life science philosophers. Should we continue to employ these well-established terms on the understanding of this teleological unburdening, or should we introduce new terms of art? In either case we risk being misunderstood, or worse. The $L = -\Lambda$ correspondence does not resolve the *proximate* question of how a biological control system realises a given set-point (although we have seen in several examples how this can be mechanistically settled: briefly, set-points can be construed as compounds of mechanistically valid parameters).

The insight that the light of evolutionary theory burns away teleology probably predates Darwin [30]. Yet even a cursory glance at the literature suggests that the issue refuses to die; there is not even a consensus that teleology does indeed constitute an embarrassment that is to be explained away, as some eminent philosophers continue to defend teleology as the proper basis for the life sciences [61] whilst others question the very coherence of evolutionary theory [13].

Likewise, the concept that hormones should be regarded as mediators of life history trade-offs has been steadily gaining ground [24, 25, 43, 101]. This concept can itself be viewed as a consequence of the more general proposition that hormones are tokens of $\nabla\Lambda$, our notation for evolutionary pressure. Even if one accepts the more general proposition, one may question whether it leads to falsifiable hypotheses, over and beyond merely being an interesting idea that ties much of organismal biology together. The crux is whether Λ is observable. In principle, it is, although we would have to be able to rerun an evolutionary process multiple times, at the population level over many generations, whilst also being able to monitor the physiological state variables in a sufficiently large sample of individuals, with sufficient resolution in time, over their life times. These are tall orders, presenting immense technical difficulties. Nonetheless, we may hope that the envisaged experiment will someday become feasible, for instance in microbial ecologies, where technologies to monitor h-states in real time have been rapidly improving [49].

❧

The terms *adaptation* and *adaptability* are generally understood to have distinct (if related) meanings on various time scales: encompassing phenomena lasting less than a second to several hours, where the organism responds to environmental challenges, but also changes on a time scale comparable to the life cycle (e.g., epigenetic changes) and finally adaptation in the evolutionary sense, where genomes are dispatched into a harsh and hostile world with increased chances of propagation over transgenerational time scales. If evolutionary adaptation ($\sim \Lambda$) is the creation of a repertoire of physiological behaviour, shorter-term adaptation ($\sim L$) can be viewed as the implementation of that repertoire. If so, the $L = -\Lambda$ principle suggests that *adaptation* and *adaptability* are not used in disparate senses between these hugely different time scales, after all.

❧

The theoretical claim staked by the present monograph is that marginal fitness can be explicitly represented as a functional of a projection of the h-state, as it follows its dynamical trajectory over the life of the organism. This generalises the perspective of the renewal integral. The original motivation is the idea of measuring fitness by accumulating reproductive output (e.g. number of eggs laid) over time. The concept was then widened to accommodate other parental investments, inclusive fitness investments into more distant kin, and so on [53, 89]. The present claim is that a 'pre-fitness' $\varrho^*(\boldsymbol{\xi}, a)$ can be assigned to the h-state $\boldsymbol{\xi}$ attained at age a, and this pre-fitness can be construed as an integral over time, such that integration over the entire lifetime turns the pre-fitness into final fitness.

These ideas keep with the received wisdom that survival-weighted fluxes of matter and energy towards reproductive success will be closely associated with fitness, and that all other aspects of the h-state ultimately affect fitness via their impact on the investments accounted for by the renewal integral. The present treatment contributes to this the liberty to deal with the relevant portion of the h-state directly, without *necessarily* having to worry over the details of genomics and population genetics, or the difficult question of how to account for some of the rather diffuse effects an organism may exert on its environment. These detailed questions are not to be disparaged, but they may be evaded or deferred whenever our primary interest is the link between a physiological subsystem (a portion of interest of the h-state) and its effects on fitness.

❧

The hope, in short, is to bolster the prospects for the 'hormones betoken $\nabla\Lambda$' programme. Limitations on what is technologically achievable represent one significant hurdle, and we may envisage that, once this hurdle is overcome, the central tenet of integrative (neuro)endocrinology should soon become standard textbook fare.

❧

We have restricted our attention to 'classic' homeostasis, in which the set-points can be assumed to be more or less constant over the relevant portion of the organism's life span. However, the present approach is equally well-appointed vis-à-vis time-varying set-points, inasmuch as the apparatus that leads from evolution to Λ, thence to L, and finally to the dynamics, Eq. (3.25), is fully applicable to such time-varying set-points.

Nevertheless, non-constant set-points do pose substantial technical challenges. The least problematic are life-stage transitions, for instance, from a growing stage to a reproductive stage, or changes in morphology and physiology that occur as the organism switches to a different feeding mode, type of prey, etc. In many cases, such life histories can be described as piece-wise constant set-points, with regulation during each life stage corresponding to a particular constellation of classic homeostatic loops. The transitions are then modelled as sudden jumps, i.e., discontinuities in the set-point value as a function of time, and the modelling task is directed primarily towards the timing of these jumps [12].

By contrast, *allostasis* can be characterised as a continuous adjustment of set-points to prevailing conditions. If $\widehat{\xi}(t) \in \mathcal{X}$ is the vector collecting the set-points evolving in time, the task is to specify a function from h-space to $\mathcal{X}$ that gives $\widehat{\xi}(t)$. To capture this relevant portion of h-space, additional $\mathcal{X}\sim$ and $\mathcal{Z}\sim$ variables might be required, as well as additional elements of $\boldsymbol{u}$.

❧

If information about the environment (the ambient medium in which the organism resides) is to be part of the determination of control outputs, the control system must be capable of gathering this information (cf. Fig. 2.1). In physiological terms, this task falls to sensory organs, which convert external stimuli (energy and matter in the environment, typically in tiny amounts) into internal stimuli such as action potentials and first messengers.

No sensory organ may be able to serve this role, perhaps because the physical intermediate of the signal has fallen below a detection threshold, or resides unfeasibly remote in space. Remoteness in time, too, often plays a key role: the relevant h-state variables are *bundles of histories* (in a-time) of environmental variables. These h-state variables can be represented by $\mathcal{X}\sim$variables that 'track' histories (standard $\mathcal{X}\sim$variables can also be said to do this, since indeed any state variable is by definition an equivalence set of histories [64]; however, here we refer to arrangements that are dedicated to capturing past events over and above what is already represented in the 'regular' physiological state). Let us use the term *engram variables* to refer to such special $\mathcal{X}\sim$variables.

Engram variables account for the nervous system's capability for *sensitisation* and *habituation* [41]. Moreover, 'memory' as commonly understood [41] is represented by engram variables. Another example is the 'trace' left in the adaptive immune system by previous exposure to pathogens [65].

❧

It is not clear if the discussion in the literature surrounding allostasis focusses properly on the adjustment or adaptation of set-points, which must accordingly be treated as dynamic variables — either as engram variables or as 'biological clock'-type variables. Certainly there can be little debate as to whether set-point shifts occur when the organism passes between the major stages of its life history (as discussed in Section 1.1).

However, not all dynamic behaviour entails an alteration of set-point values. The central idea in Chapter 6 was that of a hierarchy of set-points, expressed by Eq. (6.15). The ketogenic (lipostatic) minimum is quite shallow and readily abandoned when the caloric intake exceeds or falls short of demand for prolonged periods of time, which is precisely what allows the glucogenic and nitrogen set-points to be protected under these conditions. Likewise, the glucogenic set-points (glycogen stores) yield more easily than the nitrogen set-point, which prolongs the body's ability to perform core physiological functions even during severe starvation. Nonetheless, if conditions are favourable, the weak pressure set up by $\partial^2 L/\partial x_\mathrm{K}^2$ suffices to drive the ketoplastic reserves to the setpoint defined by $\partial^2 L/\partial x_\mathrm{K}^2 = 0$.

What do we mean precisely when we say that set-points are abandoned under these 'allostatic' challenges? We have defined set-points as minima of L, and these do not change: the set-point as such retains its value (conversely, when we talk of shifts of set-points, we do mean a change in L such that its minima are altered). To conclude that these set-points are no actual set-points after all amounts to a fallacy of the 'no true Scotsman' variety: a set-point would then only be a true set-point if the second derivative of L about the set-point were effectively infinite.

Should there perhaps be a threshold for $\partial^2 L/\partial x^2$ below which we have allostasis ('soft' homeostasis) and above which we have homeostasis (tight, narrow-range control)? Or does Nature draw the line for us, assigning some mediators a role in allostasis ('soft' homeostasis, adaptation to adversity) and others in classic (tight) homeostasis? McEwen and Wingfield [54] hint at both:

> [A]llostasis can be defined as the active process of maintaining/re-establishing homeostasis, when one defines homeostasis as those aspects

> of physiology (pH, oxygen tension, body temperature for homeotherms) that maintain life. In that context, allostasis refers to the ability of the body to produce hormones (such as cortisol, adrenalin) and other mediators (e.g. cytokines, parasympathetic activity) that help an animal adapt to a new situation/challenge [...] [S]ome colleagues prefer to use homeostasis to mean the same as allostasis, in which case homeostasis refers to those aspects of physiology that keep us alive as well as those aspects of physiology that help us adapt. The advocates of "homeostasis only" say that this term has long been used to mean both things and to change it would be confusing, or worse, but they do not consider the following: [a] cardinal feature of allostasis is that there are hugely different levels of activity of those mediators involved in adaptation — e.g. elevated heart rate, blood pressure, cortisol, or inflammatory cytokines — that may be needed in the short term to help us adapt, or which may occur chronically and lead to disease (e.g. hypertension, Cushing's disease, certain forms of depression, arthritis, metabolic syndrome). There is a key difference here: in contrast to the mediators (of allostasis) that actively promote adaptation, those features that maintain life [...] are ones that operate in a narrower range and do not change in order to help us adapt; i.e., they are not the mediators of change.

Accordingly, we might perhaps designate factors such as insulin and leptin (classic) homeostatic mediators, as opposed to, say, inflammatory cytokines or cortisol, which would be allostatic mediators. Yet these mediators often appear to fulfill both roles; cf. Example 14, for instance. Moreover, McEwen and Wingfield [54] present caloric over- or under-supply in the diet as their prime example of allostatic (over)load, which, as we have seen in Chapter 6 can be largely analysed and explained without necessarily assuming that the minima of L shift, i.e., 'classic' homeostasis by their definitions.

Of course such alterations in L may well occur in reality in response to prolonged dietary shifts, and, moreover, McEwen and Wingfield have a point when they claim that the 'homeostasis only' point of view (*sensu stricto*, as they conceive of it) fails to incorporate anticipatory changes such as initiation of lactation prior to parental care, migration to a breeding ground, or reproductive development prior to seasonal breeding [54]. For the theory presented in the present monograph, time-dependent (or better: life stage- and general physiological state-dependence) is not grudgingly tacked on as an afterthought to accommodate the above objections, but is part of the very foundations, as close study of the development in Section 7.3 will reveal.

What's in a name? McEwen and Wingfield [54] ask. Should 'homeostasis' be restricted to those cases were L itself is a-invariant and 'allostasis' to cases where L does vary (although this would still leave some of their 'allostasis' in the wrong pigeonhole)? The life sciences are in desperate need of terminology designating deep mathematical concepts rather than making superficial differences appear to be paramount. It is ultimately up to the community to decide upon standard usage.

❧

We have made frequent use of the distinction between $\mathcal{X}\sim$ and $\mathcal{Z}\sim$variables, whose boundary should be thought of as flexible — subject to preference, aims, available data, and so on. This mobility does not pose a conceptual problem, since $\mathcal{X}\sim$ and $\mathcal{Z}\sim$variables both fall under the h-state. Still, it may be objected that, in its formal aspects, the theory relies heavily on a distinction that is not clearly drawn. The conceptual problem is just this: as modellers, we take decisions on the spatiotemporal resolution (fine- versus coarse-graining, lumping, averaging, and so on) of the $\mathcal{X}\sim$component. Once laid down, these decisions also constrain the spatiotemporal resolution of the $\mathcal{Z}\sim$component. The dynamics of the regulatory component is determined by the behaviour of its actuators, i.e., at the level where it connects to the $\mathcal{X}\sim$component.

We have discussed various ways to work around this problem. For instance, instead of having the flux commanded by a hormone as the actuator, we can have the concentration of the hormone as $\mathcal{X}\sim$variable and its secretion rate as $\mathcal{Z}\sim$variable. If this does not fix the problem, we can go up the 'efferent chain' one step further. Ultimately however, we feel dissatisfied as we are addressing a fundamental problem with *ad hoc* fixes. The $\mathcal{X}\sim$ and $\mathcal{Z}\sim$variables should be treated on an equal footing, being part of the h-state.

❧

The problem of appropriate coarse-graining also affects our ability to predict pleiotropic suites on the basis of a coupling graph. In Example 14, we were able to postulate detailed correspondences with hormones, Eq. (14), and obtain a good agreement with the suites of these hormones via the coupling graph, Fig. 4.3. If we attempt to perform a similar exercise with the coupling graph depicted in Fig. 6.1, we run into trouble. The culprit is excessive lumping. The reserve variables represent not only storage-type reserves in the form of cellular inclusions of biopolymers in dedicated cells, but also nutrients carried by the bloodstream. There is a similar lumping in the $\mathcal{Z}\sim$variables, which each represent localised fluxes accumulated over the entire organism; notably the conversion of ketogenic carbon to glucogenic carbon, which has both an assimilatory component and a dissimilatory component. If we explicitly account for the bloodstream compartment in our model of the physiological component, we find that we can readily trace the regulatory pressures and associated hormones — after all, we already did the work in Example 14.[†]

From the vantage point of standard modelling practice, however, the reasons for lumping are perfectly sound: the majority of the reserves is present in the form of long-term storage, with the amount present in the blood being an extremely small fraction of the total reserves, consistent with the primary role of nutrient homeostasis in the blood plasma, which is to provide steady and reliable concentrations of substrates to the body's tissues (cf. the discussion in Section 2.1). Moreover, the analysis is concerned with starvation, refeeding, overfeeding, and somatic growth, processes associated with a time scale that is slow relative to fluctuations of the blood plasma titres (which are dominated by the diurnal cycle as well as variations in the availability of food), so it seems perfectly reasonable to assume that the blood levels remain quasi-constant at their long-time average, and that the fluxes ($\mathcal{Z}\sim$variables) in the model are similarly averaged over a window of e.g. a 10-hour time span.

If we compare a model in which the hub pools are explicitly represented (e.g. as portrayed schematically in Figs. 1.1, 4.2, and 4.3) with the lumped model (Fig. 6.1), we see that the regulatory pressures associated with the stores are transmitted *via* the levels of small molecules dissolved in the blood. Perhaps the long-term average of the latter must change in order to transmit these regulatory pressures over the long-term (i.e., the time scale associated with the reserves), and while this may occur, it is not necessary: it already suffices if there is a change in the waveform *shape* of the short-term (e.g., hourly, daily) fluctuations, while the waveform *average* remains constant. This comes about because, if $\wr \cdot \wr$ denotes the smoothing operation over an appropriate short time scale, then $\wr f(\boldsymbol{\xi})\wr \neq f(\wr\boldsymbol{\xi}\wr)$, unless, generically, the function f is linear. Not only are relationships in biological systems typically non-linear, but the nonlinearity often proves to be essential to system function.

There are two take-home messages. The first is an incompatibility between the claims for (i) L-plus-$\boldsymbol{f}$ as general modelling strategy and (ii) hormones as tokens of the gradient of L. At an intermediate timescale, where short-term interventions via some mediators meet long-term interventions of others, we have to relinquish the literal tokens interpretation. The situation is further complicated by the long-term development of the 'short term' hormone levels. Conversely, if we wish to retain the 'tokens' interpretation, we have to work on the shorter time scale, or else be much more careful in how we forge the link to longer-term dynamics.

[†]By regulatory pressures we mean terms that go as ∇L. If we countenance the more ambitious claim that $L = -\Lambda$, these would equally count as evolutionary pressures.

This requires *homogenisation*, a systematic method of deriving the long-term appearance of dynamics given on a shorter time scale; in view of the $\wr f(\xi)\wr \neq f(\wr\xi\wr)$ caveat noted above, this is not a trivial exercise [87].

The upshot is that we should adapt our model to the *milieu intérieur* per se, if we want hormones-as-tokens to hold good. But this is not altogether surprising in view of our account of the origins of homeostasis, Section 2.1. Similar considerations apply when we wish to connect the token interpretation to the fitness landscape for other types of biological signals such as second messengers, transcription factors, cell-adhesion molecules, surface-to-surface interactions, and neurotransmitters.

❧

The coupling graph in Fig. 4.3 led us to consider the walks-in-graphs method that can be used to determine the signs of a given hormone's pleiotropic effects, provided we are prepared to identify that hormone with a particular homeostatic pressure (i.e., ∇L-type term). This gave us some insight into the (neuro)endocrine control of nutrient disposition and organismal energy budget management. However, numerous important questions remain unanswered. One such question is the functional significance of the hypothalamic-pituitary system, which can be regarded as a centralised system that coordinates energetic expenditures with investment into body mass and reproduction, and adjusts these various fluxes to the animal's sleep/wake cycle [16].

❧

Regulation comes with a fitness cost. Building blocks and energy are diverted to the development and maintenance of regulatory infrastructure, and are as such divested from other destinations such as somatic growth and reproduction.[‡] We have frequently alluded to this principle; for instance, the argument surrounding Eq. (7.12) is predicated on the cost of regulation.

Yet we have not come to terms with this cost in an entirely satisfactory manner, and the basic difficulty is again that the distinction between $\mathcal{X}\sim$ and $\mathcal{Z}\sim$variables is one of great convenience, but one that will eventually have to be dissolved.

The intuitive question of how much regulation a unit of fitness buys is not easily put on a rigorous, general footing. Consider the concentration

[‡]It has been suggested that the notion of trade-offs should be interpreted with due care: a trade-off, properly speaking, is a 'hypothesis concerning the cause of a negative trait association' [101]. On the other hand, whenever trade-offs can be directly linked to conservation principles, such methodological niceness is too much of a good thing.

of nerve ganglia into the beginnings of a central nervous system. Both in terms of development and maintenance, this represents a considerable investment. At the same time, centralisation of the neuro-endocrine system represents a pre-adaptation for further innovations that ultimately enables the organism to exert a much greater influence over both *milieu intérieur* and *extérieur*, although, of course, the adaptive benefit must be present at every step, as pre-adaptations are not evolutionarily favoured for the benefits they will not bring before countless generations have come and gone.

Moreover, we are grossly oversimplifying when we talk of regulatory components evolving to steer physiological components. In reality, evolutionary change becomes available to the latter as soon as regulatory structures arise. The physiological and regulatory components co-evolve. On an intuitive level, these are truisms, but even so, the best direction for theory to take is not obvious. M. J. West-Eberhard's admirable *Developmental Plasticity and Evolution* [89] can be viewed as a standard-bearer of the field, representing both its strengths and weaknesses. We know a great deal about how organisms function throughout their development and have strong intuitions regarding gene-to-trait, evo-devo and co-evo-devo, etc., but the task remains of lifting out of this morass of *faits divers* a formalism that is explicit where it counts, sufficiently systematic and general, and rigorous.[††]

❧

Closing on a more speculative note, let us consider how far the idea

$$\text{signals in biological systems} \equiv \\ \text{gradients of } L \propto \Lambda = \\ \text{fitness landscape}$$

could be taken. There is an obvious danger of over-egging the pudding: we stray into the realm of vacuous metaphor when we fail to attend to matters such as observability and other niceties of the sorts we have been discussing. Let us suppose, however, that such problems can be overcome, and consider increasingly unfettered avenues for our imagination.

First, our examples have been of the bread-and-butter variety. This was deliberate, in the sense that too much audacity in the models themselves would distract from the main thrust of the arguments. As a result we have been taking the 'scene of the action' (the $\mathcal{X}$~component) as a

[††]The literature abounds with claims of having achieved just such a grand follow-up to the modern synthesis, but this genre is long on straw-man arguments and short on any sort of relevant mathematics.

fixed given and enquired what types of regulation might be required. We should like to be able to describe how the '$\mathcal{X}$~component' evolves as the advent of regulation opens up novel ways of living lives.

While we have focused here on mammalian-based examples (so as not to court controversy more than strictly necessary), the principles apply equally to other kingdoms of life, notably the prokaryotes [62]. From the perspective of the present monograph, there are striking points of commonality between the evolution of multicellularity in aggregates of eukaryotic cells and the origin of eukaryotic cells as endosymbiontic organismal harmonies between prokaryotic partners. Similarly, symbiontic assemblies at the level of meta-individuals (microbial mats, lichens, nests of social insects) should be amenable to an analysis closely following the lines we have set out, although the fitness landscape risks being more 'diffuse,' as can be seen by closely following the construction in Sections 7.2 and 7.3 and considering the alterations required for this application.

The kind of regulation we have in mind here arises as a product of, and is a pre-condition for, the evolution of multicellularity[‡‡] out of multicellular colonies, which involves specialisation and differentiation, e.g. groups of cells taking on distinct metabolic functions, such as biochemical conversions that are incompatible with other processes in terms of the physico-chemical environment they require, reserve polymer storage functions, integumental and locomotory functions, and so on. A central compartment of interstitial fluid provides the beginnings of a *milieu intérieur* and is at the same time the conduit for first-messenger molecules.

[‡‡]True multicellularity could be said to have been attained when propagules consisting of a single cell (or perhaps relatively few cells) ensure distinct 'germ line' reproduction, as opposed to budding-off or stolon-type reproduction. These innovations would be expected to arise somewhat later than metabolic division of labour.

Bibliography

[1] W. J. van Aardt. Quantitative aspects of the water balance in *Lymnaea stagnalis* (L.). *Neth. J. Zool.*, 18:253–312, 1968.

[2] J. Alba-Roth, O. A. Müller, J. Schopohl, and K. von Werder. Arginine stimulates growth hormone secretion by suppressing endogenous somatostatin secretion. *J. Clin. Endocrinol. Metab.*, 67:1186–1189, 1988.

[3] C. Allen. Teleological notions in biology. *In: The Stanford Encyclopedia of Philosophy.* `https://plato.stanford.edu/entries/teleology-biology/`, 2009.

[4] J. H. Andrews. *Comparative Ecology of Microorganisms and Macroorganisms.* Springer, Berlin, 2017.

[5] J. Atia, C. McCloskey, A. S. Shmygol, H. A. van den Berg, and A. M. Blanks. Reconstruction of cell surface densities of ion pumps, exchangers, and channels from mrna expression, conductance kinetics, whole-cell calcium, and current-clamp voltage recordings, with an application to human uterine smooth muscle cells. *PLoS Comput. Biol.*, 12:e1004828, 2016.

[6] J.-P. Aubin and A. Cellina. *Differential Inclusions: Set-Valued Maps and Viability Theory.* Springer, Berlin, 1984.

[7] F. Bauer and C. Fike. Norms and exclusion theorems. *Numer. Math.*, 2:137–141, 1960.

[8] D. J. Bell and D. H. Jacobson. *Singular Optimal Control Problems.* Academic Press, London, 1975.

[9] H. A. van den Berg. Modelling the metabolic versatility of a microbial trichome. *Bull. Math. Biol.*, 60:131–150, 1998.

[10] H. A. van den Berg. Propagation of permanent perturbations in food chains and food webs. *Ecol. Modell.*, 107:225–235, 1998.

[11] H. A. van den Berg. *Mathematical Models of Biological Systems*. Oxford University Press, Oxford, 2011.

[12] H. A. van den Berg. *Evolutionary Dynamics: The Mathematics of Genes and Traits*. Institute of Physics, London, 2015.

[13] H. A. van den Berg. Darwin endures, despite disparagement. *Science Progress*, 101:32–51, 2018.

[14] H. A. van den Berg, Yu. N. Kiselev, and M. V. Orlov. Homeostatic regulation in physiological systems: A versatile Ansatz. *Math. Biosci.*, 268:92–101, 2015.

[15] H. A. van den Berg, M. Orlov, and Y. N. Kiselev. The Malthusian parameter in microbial ecology and evolution: An optimal control treatment. *Comp. Math. Model.*, 19:406–428, 2008.

[16] C. G. D. Brook and N. J. Marshall. *Essential Endocrinology*. Blackwell Science, Oxford, 2001.

[17] R. Casey, H. de Jong, and J.-L. Gouzé. Piecewise-linear models of genetic regulatory networks: Equilibria and their stability. *J. Math. Biol.*, 52:27–56, 2006.

[18] A. Chatzitomaris, R. Hoermann, J. E. Midgley, S. Hering, A. Urban, B. Dietrich, A. Abood, H. H. Klein, and J. W. Dietrich. Thyroid allostasis-adaptive responses of thyrotropic feedback control to conditions of strain, stress, and developmental programming. *Front. Endocrinol.*, 8:163, 2017.

[19] B. Corentin, A. Gupta, and M. Khammash. Antithetic Integral Feedback ensures robust adaptation in noisy biomolecular networks. In *eprint arXiv:1410.6064v7*, 2016.

[20] M. di Bernardo, C. Budd, A. R. Champneys, and P. Kowalczyk. *Piecewise-smooth Dynamical Systems: Theory and Applications*. Springer, Berlin, 2008.

[21] J.-P. Draye and J. Vamecq. The gluconeogenicity of fatty acids in mammals. *TIBS*, 14(478–479), 1989.

[22] Irving M. Faust, Patricia R. Johnson, Judith S. Stern, and Jules Hirsch. Diet-induced adipocyte number increase in adult rats: A new model of obesity. *Am. J. Physiol.*, 235:279–286, 1978.

[23] A. F. Filippov. *Differential Equations with Discontinuous Righthand Sides*. Kluwer Academic, Dordrecht, 1988.

[24] C. E. Finch and M. R. Rose. Hormones and the physiological architecture of life history evolution. *Quart. Rev. Biol.*, 70:1–52, 1995.

[25] T. Flatt, M.-P. Tu, and M. Tatar. Hormonal pleiotropy and the juvenile hormone regulation of *Drosophila* development and life history. *BioEssays*, 27:999–1010, 2005.

[26] K. N. Frayn. *Metabolic Regulation: A Human Perspective.* W. B. Saunders, Philadelphia, 2003.

[27] P. Godfrey-Smith. A modern history theory of functions. *Noûs*, 28:344–362, 1994.

[28] F. Götze and A. Tikhimorov. Rate of convergence in probability to the Marchenko-Pastur law. *Bernoulli*, 10:503–548, 2004.

[29] S. Green. Revisiting generality in biology: Systems biology and the quest for design principles. *Biol. Philos.*, 30:629–652, 2015.

[30] M. Grene and D. Depew. *The Philosophy of Biology: An Episodic History.* Cambridge University Press, Cambridge, 2008.

[31] D. J. Griffiths and D. F. Schroeter. *Introduction to Quantum Mechanics.* Cambridge University Press, Cambridge, 2018.

[32] A. C. Guyton. The surprising kidney-fluid mechanism for pressure control — its infinite gain! *Hypertension*, 16:725–730, 1990.

[33] W. D. Hamilton. The moulding of senescence by natural selection. *J. Theor. Biol.*, 12:12–45, 1966.

[34] D. B. Hausman, M. DiGirolamo, T. J. Bartness, G. J. Hausman, and R. J. Martin. The biology of white adipocyte proliferation. *Obesity Rev.*, 2:239–254, 2001.

[35] A. K. Hewson, L. Y. C. Tung, D. W. Connell, L. Tookman, and S. L. Dickson. The rat arcuate nucleus integrates peripheral signals provided by leptin, insulin, and a ghrelin mimetic. *Diabetes*, 51:3412–3419, 2002.

[36] M. A. H. Hoskins and M. Aleksiuk. Effects of temperature on the kinetics of malate dehydrogenase from a cold climate reptile, *Thamnophis sirtalis parietalis*. *Comp. Biochem. Physiol. B*, 45:343–346, 1973.

[37] J.-Y. Hsu, S. Crawley, M. Chen, D. A. Ayupova, D. A. Lindhout, J. Higbee, A. Kutach, W. Joo, Z. Gao, D. Fu, C. To, K. Mondal, B. Li, A. Kekatpure, M. Wang, T. Laird, G. Horner, J. Chan, M. McEntee, M. Lopez, D. Lakshminarasimhan, A. White, S.-P. Wang, J. Yao, J. Yie, H. Matern, M. Solloway, R. Haldankar, T. Parsons, J. Tang, W. D. Shen, Y. Alice Chen, H. Tian, and B. B. Allan. Non-homeostatic body weight regulation through a brainstem-restricted receptor for GDF15. *Nature*, 550:255–259, 2017.

[38] J. Jacobs. Teleology and reduction in biology. *Biol. Philos.*, 1:389–399, 1986.

[39] O. L. R. Jacobs. *Introduction to Control Theory*. Oxford Science Publications, Oxford, 1993.

[40] S. Kamohara, R. Burcelin, J. L. Halaas, J. M. Friedman, and M. J. Charron. Acute stimulation of glucose metabolism in mice by leptin treatment. *Nature*, 389:374–377, 1997.

[41] E. Kandel, J. Schwartz, T. Jessell, S. Siegelbaum, and A. J. Hudspeth. *Principles of Neural Science*. McGraw-Hill, New York, Fifth edition, 2012.

[42] J. Keener and J. Sneyd. *Mathematical Physiology*. Springer, Berlin, 1998.

[43] E. D. Ketterson and J. W. Atwell. *Snowbird: Integrative Biology and Evolutionary Diversity in the Junco*. University of Chicago Press, Chicago, 2016.

[44] M. S. Kim, C. J. Small, S. A Stanley, D. G. A. Morgan, L. J. Seal, W. M. Kong, C. M. B. Edwards, S. Abusnana, D. Sunter, M. A. Ghatei, and S. R. Bloom. The central melanocortin system affects the hypothalamo-pituitary thyroid axis and may mediate the effect of leptin. *J. Clin. Invest.*, 105:1005–1011, 2000.

[45] D. L. Kirk. A twelve-step program for evolving multicellularity and a division of labor. *BioEssays*, 27:299–310, 2005.

[46] D. S. Kompala, D. Ramkrishna, and G. T. Tsao. Cybernetic modelling of microbial growth on multiple substrates. *Biotechnol. Bioeng.*, 26:1272–1281, 1984.

[47] S. A. L. M. Kooijman. *Dynamic Energy and Mass Budgets in Biological Systems*. Cambridge, 2003.

[48] K. Kumar, R. A. Mella-Herrera, and J. W. Golden. Cyanobacterial heterocysts. *Cold Spring Harb. Perspect. Biol.*, 2:a000315, 2009.

[49] E. Larrainzar, F. O'Gara, and J. P. Morrissey. Applications of autofluorescent proteins for in situ studies in microbial ecology. *Annu. Rev. Microbiol.*, 59:257–277, 2005.

[50] E. R. Leadbetter and J. S. Poindexter. *Bacteria in Nature Volume 3: Structure, Physiology, and Genetic Adaptability*. Plenum Press, New York, 1989.

[51] R. E. Lenski, M. R. Rose, S. C. Simpson, and S. C. Tadler. Long-term experimental evolution in *Escherichia coli*. I. Adaptation and divergence during 2,000 generations. *Amer. Nat.*, 138:1315–1341, 1991.

[52] V. A. Marčenko and L. A. Pastur. The distribution of eigenvalues in certain sets of random matrices. *Mat. Sb.*, 72:507–536, 1967.

[53] J. Maynard Smith. *Evolutionary Genetics*. Oxford University Press, Oxford, 1999.

[54] B. S. McEwen and J. C. Wingfield. What's in a name? Integrating homeostasis, allostasis and stress. *Horm. Behav.*, 57:105–111, 2010.

[55] T. J. Merimee, D. Rabinowitz, and S. E. Fineberg. Arginine-initiated release of human growth hormone. Factors modifying the response in normal man. *N. Engl. J. Med.*, 280:1434–1438, 1969.

[56] J. A. J. Metz. Fitness. In S. E. Jørgensen and B. D. Fath, editors, *Encyclopedia of Ecology*, pages 1599–1612. Elsevier, Oxford, 2008.

[57] J. A. J. Metz, S. A. H. Meszena, F. J. A. Jacobs, and J.S. van Heerwaarden. Adaptive dynamics: A geometrical study of the consequences of nearly faithful reproduction. Technical Report WP-95-099, IIASA, Laxenburg, Austria, 1995.

[58] G. Michal. *Biochemical Pathways*. Spektrum Akademischer Verlag, 1999.

[59] D. B. Miller, S. J. Snow, M. C. Schladweiler, J. E. Richards, A. J. Ghio, A. D. Ledbetter, and U. P. Kodavanti. Acute ozone-induced pulmonary and systemic metabolic effects are diminished in adrenalectomized rats. *Toxicol. Sci.*, 150:312–322, 2016.

[60] M. G. Myers and D. P. Olson. Central nervous system control of metabolism. *Nature*, 491:357–363, 2012.

[61] T. Nagel. *Mind and Cosmos: Why the Materialist Neo-Darwinian Conception of Nature is Almost Certainly False*. Oxford University Press, Oxford, 2012.

[62] O. A. Nev, O. A. Nev, and H. A. van den Berg. Optimal management of nutrient reserves in microorganisms under time-varying environmental conditions. *J. Theor. Biol.*, 429:124–141, 2017.

[63] J. W. Osborn. Hypothesis: Set-points and long-term control of arterial pressure. A theoretical argument for a long-term arterial pressure control system in the brain rather than the kidney. *Clin. Exp. Pharmacol. Physiol.*, 32:384–393, 2005.

[64] L. Padulo and M. A. Arbib. *System Theory*. Saunders, Philadelphia, 1974.

[65] P. Parham. *The Immune System*. Garland, New York, 2014.

[66] R. Pattaranit. *Glucose-Sensing in the Hypothalamic Arcuate Nucleus: Electrophysiological and Mathematical Studies*. PhD thesis, University of Warwick, Coventry, 2009.

[67] M. E. Peterson, R. Eisenthal, M. J. Danson, A. Spence, and R. M. Daniel. A new intrinsic thermal parameter for enzymes reveals true temperature optima. *J. Biol. Chem.*, 279:20717–20722, 2004.

[68] M. Pfaundler. Über die energetische Flächenregel. *Pflügers Archiv*, 188:273–280, 1921.

[69] E. Pinch. *Optimal Control and the Calculus of Variations*. Oxford Science Publications, Oxford, 1995.

[70] M. N. Price, K. M. Wetmore, R. J. Waters, M. Callaghan, J. Ray, H. Liu, J. V. Kuehl, R. A. Melnyk, J. S. Lamson, Y. Suh, H. K. Carlson, Z. Esquivel, H. Sadeeshkumar, R. Chakraborty, G. M. Zane, B. E. Rubin, J. D. Wall, A. Visel, J. Bristow, M. J. Blow, A. P. Arkin, and A. M. Deutschbauer. Mutant phenotypes for thousands of bacterial genes of unknown function. *Nature*, 557:503–509, 2018.

[71] A. Pütter. Studien über physiologische Ähnlichkeit VI. Wachstumsähnlichkeiten. *Pflügers Archiv*, 180:298–340, 1920.

[72] R. A. Raff. *The Shape of Life: Genes, Development, and the Evolution of Animal Form.* University of Chicago Press, Chicago, 1996.

[73] C. Regnault, M. Usal, S. Veyrenc, K. Couturier, C. Batandier, A. L. Bulteau, D. Lejon, A. Sapin, B. Combourieu, M. Chetiveaux, C. Le May, T. Lafond, M. Raveton, and S. Reynaud. Unexpected metabolic disorders induced by endocrine disruptors in *Xenopus tropicalis* provide new lead for understanding amphibian decline. *Proc. Natl. Acad. Sci.*, 115:E4416–E4425, 2018.

[74] M. R. Riddle, A. C. Aspiras, K. Gaudenz, R. Peuß, J. Y. Sung, B. Martineau, M. Peavey, A. C. Box, J. A. Tabin, S. McGaugh, R. Borowsky, C. J. Tabin, and N. Rohner. Insulin resistance in cavefish as an adaptation to a nutrient-limited environment. *Nature*, 555:647–651, 2018.

[75] D. E. Rozen, J. A. G. M. de Visser, and P. J. Gerrish. Fitness effects of fixed beneficial mutations in microbial populations. *Curr. Biol.*, 12:1040–1045, 2002.

[76] N. B. Ruderman. Muscle amino acid metabolism and gluconeogenesis. *Annu. Rev. Med.*, 26:245–258, 1975.

[77] J. G. Salway. *Metabolism at a Glance.* Blackwell Science, Oxford, 2004.

[78] K. Schmidt-Nielsen. *Animal Physiology: Adaptation and Environment.* Cambridge University Press, Cambridge, 1975.

[79] M. W. Schwartz, S. C. Woods, D. Porte, Jr., R. J. Seeley, and D. G. Baskin. Central nervous system control of food intake. *Nature*, 404:661–671, 2000.

[80] B. Sinervo. Mechanistic analysis of natural selection and a refinement of Lack's and Williams's principles. *Am. Nat.*, 154:S26–S42, 1999.

[81] G. V. Smirnov. *Introduction to the Theory of Differential Inclusions.* American Mathematical Society, Providence, 2002.

[82] A. B. Steffens. Influence of reversible obesity on eating behaviour, blood glucose, and insulin in the rat. *Am. J. Physiol.*, 228:1738–1744, 1975.

[83] J. A. Stevens. Impulse and animal action in stoic psychology. *News. Soc. Ancient Greek. Phil.*, 204:12–1996, 1996.

[84] V. I. Utkin. *Sliding Modes in Control Optimization.* Springer, Berlin, 1992.

[85] L. von Bertalanffy. Untersuchungen über die Gesetzlichkeit des Wachstums. I. Allgemeine Grundlagen der Theorie; mathematische und physiologische Gesetzlichkeiten des Wachstums bei Wassertieren. *Arch. Entwicklungsmech.*, 131:613–652, 1934.

[86] C. H. Waddington. *The Strategy of the Genes.* George Allen & Unwin, London, 1957.

[87] Y.-F. Wang, M. Khan, and H. A. van den Berg. Interaction of fast and slow dynamics in endocrine control systems with an application to β-cell dynamics. *Math. Biosci.*, 235:8–18, 2012.

[88] T. F. Weiss. *Cellular Biophysics.* MIT Press, Cambridge, Massachusetts, 1996.

[89] M. J. West-Eberhard. *Developmental Plasticity and Evolution.* Oxford University Press, Oxford, 2003.

[90] M. Wilkinson and R. E. Brown. *An Introduction to Neuroendocrinology.* Cambridge University Press, Cambridge, Second edition, 2015.

[91] N. D. de With. Evidence for the independent regulation of specific ions in the haemolymph of *Lymnaea stagnalis* (L.). *Proc. Kon. Ned. Akad. Wet. C*, 80:144–157, 1977.

[92] N. D. de With. Oral water ingestion in the pulmonate freshwater snail, *Lymnaea stagnalis. J. Comp. Physiol. B*, 166:337–343, 1996.

[93] N. D. de With and H. A. van den Berg. Neuroendocrine control of hydromineral metabolism in molluscs, with special emphasis on the pulmonate freshwater snail, *Lymnaea stagnalis. Adv. Comp. Endocrinol.*, 1:83–99, 1992.

[94] N. D. de With, J. W. Slootstra, and R. C. van der Schors. The bioelectrical activity of the body wall of the pulmonate freshwater snail *Lymnaea stagnalis*: Effects of neurotransmitters and the sodium influx stimulating neuropeptides. *Gen. Compar. Endocrinol.*, 70:216–223, 1988.

[95] N. D. de With and R. C. van der Schors. Urine composition and kidney function in the pulmonate freshwater snail *Lymnaea stagnalis*. *Comp. Biochem. Physiol.*, 79A:99–103, 1984.

[96] M. Wyss and R. Kaddurah-Daouk. Creatine and creatinine metabolism. *Physiol. Rev.*, 80:1107–1213, 2000.

[97] X. Xia, X. Wang, Q. Li, N. Li, and J. Li. Essential amino acid enriched high-protein enteral nutrition modulates Insulin-like Growth Factor-1 system function in a rat model of trauma-hemorrhagic shock. *PLoS One*, 8:e77823, 2013.

[98] A. W. Xu, C. B. Kaelin, K. Takeda, S. Akira, M. W. Schwartz, and G. S. Barsh. PI3K integrates the action of insulin and leptin on hypothalamic neurons. *J. Clin. Invest.*, 115:951–958, 2005.

[99] T.-M. Yi, Y. Huang, M. I. Simon, and J. Doyle. Robust perfect adaptation in bacterial chemotaxis through integral feedback control. *Proc. Natl. Acad. Sci.*, 97:4649–4653, 2000.

[100] W. Yourgrau and S. Mandelstam. *Variational Principles in Dynamics and Quantum Theory*. Pitman, London, 1979.

[101] A. J. Zera and L. G. Harshman. The physiology of life history trade-offs in animals. *Ann. Rev. Ecol. Syst.*, 32:95–126, 2001.

[102] M. Zimmermann. General principles of regulation. In R. F. Schmidt and G. Thews, editors, *Human Physiology*, chapter 15, pages 324–332. Springer, Berlin, 1989.

Index

D

E

F

N

O

P

Q